Fr: Damian
Sept 12

# *What Do They Call a Fisherman?*

# *What Do They Call a Fisherman?*

## *Men, Gender, and Restructuring in the Newfoundland Fishery*

Nicole Gerarda Power

ISER Books gratefully acknowledges the assistance of the Faculty of Arts, Memorial University of Newfoundland.

**National Library of Canada Cataloguing in Publication**

Power, Nicole G. (Nicole Gerarda), 1972-
What do they call a fisherman? Men, gender, and restructuring in the Newfoundland fishery / Nicole G. Power.

(Social and economic studies ; no. 69)
Includes bibliographical references and index.
ISBN 1-894725-02-6

1. Fishers – Newfoundland and Labrador – Social conditions. 2. Fishery policy – Social aspects – Newfoundland and Labrador. 3. Sexual division of labour – Newfoundland and Labrador. 4. Masculinity - Newfoundland and Labrador. I. Title.

HC117.N4P69 2004 305.33'6392'09718 C2003-907164-2

Published by ISER Books – Faculty of Arts Publications
Institute of Social and Economic Research
Memorial University of Newfoundland
St. John's, NL A1C 5S7

Printed and bound in Canada

# *Contents*

To Haley and Joey

# *Acknowledgements*

I would like to extend my gratitude to a number of people who made this work possible. I am indebted to the Newfoundland women and men who freely gave of their time and knowledge and welcomed me into their homes and community spaces. Thanks go to Dr. Carlo Ruzza, Professor Miriam Glucksmann, and Dr. Michael Roper for their supervision, time, and comments on various drafts of my Ph.D. thesis, on which this book is based. Thanks also are due to Dr. Andrew Canessa and Professor Rosemary Crompton for their academic critiques and advice on how to translate my thesis into a book. I am grateful to Dr. Barbara Neis and Dr. Larry Felt for allowing me access to a rich database of interview transcripts collected by the Eco-Research Project, Memorial University of Newfoundland. Finally, I wish to acknowledge Charlie Conway of Memorial's Cartographic Laboratory for reproducing the map of Newfoundland, as well as the efforts and helpful critiques of the anonymous reviewers of this manuscript, and Richard Tallman's editorial assistance.

This study and book were made possible through the financial support of the Rothermere Fellowship, which supported my Ph.D. studies, and the ISER Books Post-Doctoral Fellowship, which facilitated the completion of this book. For this, I thank the Trustees of the Rothermere Fellowship and the Institute for Social and Economic Research at Memorial University of Newfoundland.

My family deserves special note for constantly encouraging me throughout my academic career. In particular, I would like to acknowledge my partner, Mark, who not only offered encouraging words and support when I became frustrated with the research and writing processes, but also became my de facto technical and editorial support. Nevertheless, any faults in this final manuscript are mine alone.

# *Abbreviations*

| | |
|---|---|
| CAFSAC | Canadian Atlantic Fisheries Scientific Advisory Committee |
| CCPFH | Canadian Council of Professional Fish Harvesters |
| DFO | Department of Fisheries and Oceans |
| EI | Employment Insurance |
| EPZs | Economic Processing Zones |
| FFAW | Fish, Food and Allied Workers |
| FFTs | factory freezer trawlers |
| FPI | Fishery Products International |
| FRCC | Fisheries Resource Conservation Council |
| GAD | Gender and Development |
| GATT | General Agreement on Tariffs and Trade |
| GPS | global positioning system |
| ICNAF | International Commission for the Northwest Atlantic Fisheries |
| IQ | individual quota |
| ITQ | individual transferable quota |
| MAD | Men and Development |
| NAFO | Northwest Atlantic Fisheries Organization |
| NAFTA | North American Free Trade Agreement |
| NCARP | Northern Cod Adjustment and Recovery Package |
| SAPs | structural adjustment policies |
| TAC | total allowable catch |
| TAGS | The Atlantic Groundfish Strategy |
| TEK | traditional ecological knowledge |
| TNCs | transnational corporations |
| UI | Unemployment Insurance |
| WID | Women in Development |

# *Making Sense of Restructuring and Gender* 1

## GENDER AND RESTRUCTURING IN THE NEWFOUNDLAND FISHERY

A growing body of literature tries to make sense of the impacts of and reactions to the most recent crisis to plague the Newfoundland fishing industry[1] and its subsequent restructuring; however, much of it is neither feminist nor concerned with gender. Analyses that exclude gender offer incomplete understandings of the processes and effects of environmental crisis and restructuring, and replicate gender and other inequalities, as well as environmental degradation. The prerequisites for sustainable and equitable communities and fair economic organization will not be discovered through exclusive academic and policy-making practices that ignore the differing and unequal positions of women and men in the fishery, or assume them to be unproblematic.

The dominant view adopted by the state is that economic restructuring, informed by a neo-liberal agenda, is the solution to ecological degradation, overcapacity, and dependency on social programs. Adjustment initiatives, accompanying state-imposed moratoriums on over-fished stocks, have thus focused on reducing debt and social spending, expanding export production, increasing regulation of access and entry, and encouraging withdrawal from the fishery with financial assistance. This restructuring is linked to globalization, although the extent of this linkage is debatable. Clearly, this direction of restructuring, at least in part, is about restricting and regulating perceived state-dependents and containing "traditional" orientations to the fishery. Nevertheless, globalization has interacted with the fishing industry in the Atlantic region in a number of ways over the last couple of decades (MacDonald and Connelly, 1990a: 133; Neis and Williams, 1997: 56-7). In a context of trade liberalization, countries successfully compete with Newfoundland's processed fish products with fish substitutes, and developing countries compete with Newfoundland for marine-based and other industries by offering cheap sources of labour, lax government regulations, and corporate tax breaks. Newfoundland is left with few alternatives to the fishery as a way to make a living, which places greater pressure on the resource. In efforts to compete in the global market, processing companies have introduced new labour processes and technologies (see Neis, 1991) and in response to the destruction of local fish stocks, companies have shifted their harvesting effort to the South to source fish globally (Sinclair, 1999: 325; Ommer, 1995: 175).

Such industrial responses and the obvious strain on marine resources should not be read as evidence of Newfoundland's position *outside* globalization. As Connelly, Li, MacDonald, and Parpart (2000: 78) remind us, peripheral regions like Atlantic Canada are not "traditional" (as modernization theorists posit), nor are they somehow "pre-capitalist" (as described by Marxists). Rather, peripheral regions are part of contemporaneous global arrangements. Neis and Williams (1997: 56) argue that a global ecological revolution is underway, "based on the transformation of nature, our productive relations to nature, the reproduction of fisheries households and communities and the dominant legal, political and ideological frameworks that govern fisheries." Guided by neo-liberal economic and New Right political philosophies, this revolution is manifested in private aquaculture projects, the transfer of Western industrial harvesting technologies with dangerous capacities, and a shift from local and subsistence to commercial exploitation.

Has this revolution triggered a "crisis of masculinity"? This is the question I attempt to answer in this book. Given the fact that the Newfoundland fishery has been an important site in which rural men earn a living, live a distinctly masculine way of life, and create meaning (Andersen and Wadel, 1972a, 1972b; Silk, 1995), it might seem safe to assume that a "crisis of fish" and the state's response to such crisis mean a "crisis of masculinity." Certainly, it means a crisis in the social reproduction of fishers' households. However, ecological, economic, and industrial changes do not inevitably produce a "feminization" of men – either in terms of identity or social position. There is evidence of feminization in parts of the industry where employment increasingly takes on the traditional characteristics of women's work – insecurity, low earnings, and deskilling. However, men and masculinity remain relatively privileged in the overall gender order. Not all men benefit in the same ways or to the same extent, and those adhering or best able to adhere to the state-sanctioned version of masculinity are clearly more advantaged. Nor have all men uncritically accepted this sanctioned version, as the title of this book – a question posed by one of the respondents in this study – reflects.

## WHY ANOTHER BOOK ABOUT MEN?

This book is about men in hard times. More specifically, it is about male inshore fishers in Newfoundland during a time of restructuring and environmental crisis. A focus on men and their economic contributions is nothing new for sociologists. Academic scholarship has traditionally assumed male breadwinners and undervalued women's work. The absence of, yet need for, women in economic development – at the levels of policy-making, research, theory, and practice – encouraged Western feminists over thirty years ago to establish the alternative Women in Development (WID) approach (Connelly et al., 2000: 56-61). The WID approach focuses on equal opportunity in order "to integrate women into the national economies of their countries" and into the hegemonic development paradigm – the modernization approach – which assesses development in terms of "the adoption of Western technologies, institutions and values" (Connelly et al., 2000: 57). The success of WID, however, is limited due to its Western biases, emphasis on practical needs, and failure to account for and address women's reproductive labour, their generalized subordination, and global inequities.

Years of reworking by Western feminists and research from the Third World have resulted in more sophisticated analyses (see Kabeer, 1995; Moser, 1993). According to Kimmel (2001: 21), the value of this work lies in its recognition that development is uneven globally and nationally *and* between women and men. Simply put, development processes and economic restructuring impact women and men in different ways because they are differently and unequally located in local and global cultures and economies. According to this Gender and Development (GAD) approach, women's status in society is deeply affected by their material conditions of life and by their position in the national, regional, and global economies. GAD also recognizes that women are deeply affected by the nature of patriarchal power in their societies at the national, community, and household levels. Moreover, women's material conditions and patriarchal authority are both defined and maintained by the accepted norms and values that define women's and men's roles and duties in a particular society (Connelly et al., 2000: 62).

Equally important, GAD's focus on gender relations – not just women – opens space to include men. The inclusion radically departs from earlier and mainstream assumptions about the invisible male subject. Men have finally become gendered! After all, gender is not just about women, and more equitable social relations depend on changes in the structural position and identities of men, as well as women (White, 1997). Despite this shift in focus, most GAD work remains preoccupied with women, and in work where men are problematized, the focus is on men – usually white, middle-class men – in northern industrialized countries (Cleaver, 2002: 1). Kimmel (2001: 21-2) argues that the invisibility of men as gendered beings in development is political and perpetuates material and ideological inequalities: "The processes that confer privilege on one group and not another group are often invisible to those upon whom that privilege is conferred. Thus . . . not having to think about gender is one of the 'patriarchal dividends' of gender inequality."

This book is situated in the emerging body of literature – a sample of which is presented throughout this chapter – that attempts to make masculinity visible and explores the lives of men in a substantially different way from traditional accounts. Much of this literature has been motivated by recent debates about the fairness of feminism, the necessity of globalized production and consumption, and the morality of diverse family forms. At risk in the new focus on men and masculinity, however, is a further marginalization of women, reflected in the insistence by some for a new approach to Men and Development (MAD) (Cleaver, 2002: 5-6). In terms of Newfoundland, the world of men participating in the harvesting sector has been well documented. According to Porter (1993: 5), this documentation reflects the intersection of the province's maritime culture and a male-dominated academy that gives primacy to men's fishing work, at the expense of women's productive and reproductive work. Given this context, why should I produce yet another piece of work about men?

The key difference between this book and much of the existing work on male fishers lies in my approach. I examine male fishers from a feminist perspective that includes insights about the distribution of power among men and between men and women. Nadel-Klein and Davis (1988: ch. 2) note the absence of gender analyses in most accounts of fishing villages in Newfoundland and elsewhere, which have been preoccupied with harvesting and other male activities. Much of the literature on development and restructuring of the Atlantic region focuses on male fishers, class, and consciousness, ignoring gender (see, for example, works by Bryant, Fairley, and Veltmeyer). In his review of this literature, Veltmeyer (1990: 83-4) notes two related trains of thought. One assumes that the expansion of capital in the fishing industry will produce a capitalist class of independent producers and proletarianize the remaining producers. In this context, fishers' struggles tend to be populist, serving the interests of small capitalist fishers. The other holds that regional capitalist development and "underdevelopment," and the resistance of independent producers, constrain the penetration of capital and the direct subsumption of labour. The result is a set of complex class relations comprised of a *semi*-proletariat and small-scale "independent" producers – more accurately understood perhaps as dependent producers, sharing proletariat concerns. Both approaches, however, tend to conceptualize class independent of household and gender relations, which proves problematic when trying to predict or understand individual behaviour – a problem noted by Marxist feminist scholarship in the 1980s. Like other feminist work, this scholarship began the task of critiquing and challenging these biases and the obscuring effect of male-centred research. Most of this feminist research on Newfoundland starts from the lives and experiences of women, and rightly so, given the historical omission of women and their misrepresentation. Only a few writers have begun the task of analyzing gender relations and patriarchy by starting from the lives of men (see, for example, Davis, 1993), and this is no easy task, especially when the men in focus hold contradictory class positions.

"Men and gender" may be getting increased attention, but not all of this literature can be described as feminist or critical (see, for example, the mythopoetic and self-help literature produced by Bly, 1990, and others). In their reviews of the literature on masculinities, Pease and Pringle (2001: 2) and Cornwall (1997: 11) note the uncritical, often self-indulgent, stance of men's studies, an overgeneralization of the commonality of experience of men, and a focus on the abstract at the expense of material condition and practice. Similarly, Brod (1994: 84-5) describes much of the early work on the subjectivity of men's experiences, which ignored social structure and patriarchy, as colluding with anti-feminist work, albeit often unintentionally. Examining how marginalized or economically vulnerable men construct masculinity runs the risk of "portraying aggressive, even misogynist, gender displays primarily as liberatory forms of resistance against class and racial oppression" (Hondagneu-Sotello and Messner, 1994: 208) or "reproducing patriarchal consciousness" (Coltrane, 1994: 55).

In order to distance myself from and in anticipation of such problems,[2] I explore the relative positions and condition of men and women in the Newfoundland fishery – in addition to the cultural and ideological bases of behaviour and identity – and include in my data interviews with women. Reiterating arguments found elsewhere (Brod, 1994; Bird, 1996), Hondagneu-Sotello and Messner (1994: 215-16) conclude that it is necessary to "keep women's experience of gender oppression as close to the center of analysis as possible" in order to avert interpretations of the construction of male identities and behaviour *only* as forms of resistance to marginalization, economic or other. Similarly, Kaufman (1994: 152) argues that men, who lack power in larger society, may exercise substantial power relative to local women. Thus, an account of masculinity must locate men in *local* power relations relative to local women. The theorization of masculinities and men as power relations helps to avoid the tendency to treat men purely as victims and to focus solely on men's limitations, and makes visible the patriarchal dividend, or "the advantage men in general gain from the overall subordination of women" (Connell, 1995: 79). After all, the limitations men experience are often linked to those privileges they incur by virtue of being men (Clatterbaugh, 1996: 291-4).

At the same time, "[n]one of us live every moment of our lives in a state of subordination to others. And the relationships we have with people around us may be 'gender relations' in the sense that these are relationships in which gender makes a difference . . . but are in no sense merely one-dimensional power relations" (Cornwall, 1997: 10). It is a mistake to conceive of women only as victims and men only as agents. Neysmith (2000: 14) argues that power is best understood as "a social enactment . . . not an entity that can simply be overthrown." In other words, power is exercised, not possessed. Social relations, then, provide space for women and men to either exercise power or not, and to experience its effects. The inclusion of both women's and men's voices in my exploration of masculinity is appropriate and necessary, although not a common approach in the literature on men and masculinities.

## THEORIZING MASCULINITY

Contrary to the common-sense assumption that gender is innate, natural, timeless, and reducible to the biological, the critical new studies on men and masculinities draw on the feminist insight that gender is a dynamic social construct and a structure of social practice (see Brittan, 1989; Clatterbaugh, 1997; Coltrane, 1994; Connell, 1995, 1998; Cornwall, 1997; Kimmel, 1994, 2001; White,

1997). Connell (1995: 71) describes gender as "a way in which social practice is ordered . . . in relation to a reproductive arena. . . . Gender is social practice that constantly refers to bodies and what bodies do, it is not social practice reduced to the body. . . . Gender exists precisely to the extent that biology does *not* determine the social." People may actively engage in social practice, yet their behaviour (including how people think and interpret) is organized by gender relations. When women and men act in feminine or masculine manners, or deliberately attempt to challenge these, they are not acting without context. Rather, their social practices reflect larger configurations of gender. These configurations may reflect culturally accepted ideas about how women and men should behave or ways of distributing value, power, and resources embedded in our institutions. Masculinity, like femininity, is "a place in gender relations, the practices through which men and women engage that place in gender, and the effects of these practices in bodily experience, personality and culture" (Connell, 1995: 71). This means that we can talk about individuals thinking or behaving in masculine ways or experiencing the effects – as limitations or opportunities – of social processes and patterns that are masculine.

Of course, precisely how gender configures practice is historically and culturally specific. Not all societies have cultural accounts of masculinity, and those that do, do not delineate masculinity in the same ways. They do, however, construct masculinity in relation to femininity (Connell, 1995: 67-8). It is better, therefore, to conceive of the existence of *masculinities* rather than a unified, fixed masculinity. The new studies also recognize that just as gender organizes practice, so, too, do other structures including class, ethnicity, age, sexuality, nationality, and so forth (see Brittan, 1989; Brod and Kaufman, 1994; Connell, 1995; Harper, 1994; Hearn and Collinson, 1994; White, 1997). Not only do masculinities vary cross-culturally and over time, they also vary in any given society at any given time. The manifestations of masculinity are thus complex and plural, mediated through the social positions that men fill. Connell (1995: 76) and Connelly et al. (2000: 96) caution that recognition of the plurality of masculinities is not the same as a postmodernist reduction of differences to individual lifestyle choices. A focus on relations among men highlights the existence of social patterns, as well as hierarchies of masculinities.

Not all masculinities are equal. Each society, region, or nation has a hegemonic model of masculinity – the version of masculinity that appears natural or is most highly valued – and serves as the standard for all men (Kimmel, 2001: 22-3). Connell (1995: 77) defines hegemonic masculinity as "the configuration of gender practice which embodies the currently accepted answer to the problem of the legitimacy of patriarchy, which guarantees (or is taken to guarantee) the dominant position of men and the subordination of women." This definition implies that the subordinate position of women is central in the construction of hegemonic masculinity. To fully unpack hegemonic masculinity, we must consider the available and accepted discourses – those particular ways of talking, writing, and doing (Neysmith, 2000: 15) – at the micro and macro levels. Cornwall (1997: 11) argues that the concept of "hegemonic masculinity" enables a distinction to be made between men and masculine ways of doing, behaving, and thinking. A patriarchal society at any particular historical moment values discourses – which are ideological when they help produce and sustain systems of inequality – and practices that privilege men and the masculine or reflect social relations that confer power to men. Discourses limit and enable particular explanations and ways of thinking about and acting in society, which in turn shapes identity. Individual men, from time to time or over one's life course, may not subscribe to these ideas or may choose not to behave in these valued ways. Individual men can and do decide not to conform to gender expectations, to take up more marginal versions of masculinity, and not to accept the patriarchal benefits ascribed to them, although not without consequences. The hegemonic, then, is always contested – at the micro level where men (and women) reject conforming to the hegemonic model, and at the macro level where alternative, challenging versions of masculinity are available in the wider culture and institutionally. From such contestation the gendered configurations of practice can change. This study reveals a complex web of relations and hierarchies as various competing and complementary discourses emerge around the "traditional" and "modern" male fisher models.

## WHAT DOES GENDER HAVE TO DO WITH RESTRUCTURING?

Gender configures all social practice – this includes our economic and political arrangements. Before I examine the relationship between gender and restructuring, some definitions are in order. Restructuring is not a new phenomenon. It's really another way of talking about recent trends in development. Today the term "restructuring" is used as a catchphrase for the presumed necessary changes in local and international economic, social, environmental, and political arrangements that have taken place in the last couple of decades (Connelly et al., 2000: 65). Neis, Grzetic, and Pidgeon (2001: 8) suggest that restructuring is multi-faceted:

> Overharvesting, reduced biodiversity and pollution are examples of environmental restructuring. Industrial restructuring processes include work reorganization (deskilling and reskilling), downsizing, outsourcing, and capital flight. Political restructuring processes include trade liberalization, privatization, deregulation, changes to public services and social programs. Social restructuring refers to such processes as urbanization, demographic change, and changing community and household dynamics.

Transnational corporations (TNCs), supranational economic organizations (like the International Monetary Fund and the World Bank), and often times national and local politicians argue that restructuring is a painful but inevitable response to globalization – those recent trends of global liberalization, the internationalization of production, and the universalization of capitalism (Bakkar, 1996: 3; Neysmith, 2000: 10-11). While the historical record demonstrates transnational contact between nations, "globalization" is usually taken to refer to the recent international neo-liberal economic consensus (Pringle and Pease, 2001: 251). Nation-states have responded to such consensus with initiatives that focus on reducing state spending, corporate taxation, and regulation, and increasing exports, instituting free trade, and privatizing public services (Bakkar, 1996: 4). These policies, in turn, impact the daily lives of people, for example, in the form of changed labour processes and work conditions and increased unemployment (Connelly et al., 2000: 65-6; Leach and Winson, 1995: 344). However, I should not overstate the impact of globalization on national economies. Uncritical views conflate globalization and restructuring. Yet, national political restructuring (for example, cuts to social programs) may serve other purposes than further integration into the global economy – for example, to contain and punish perceived abusers of the welfare state (Neysmith, 2000: 9-10). Other uncritical views see restructuring as an inevitable response to global trends. Yet, nation-states and politicians continue to be powerful actors and can resist, constrain, or assist neo-liberal agendas (Christie, 2001: 108; Pease and Pringle, 2001: 9).

This fact that the nation and its politicians have disappeared in discussions on restructuring reflects the success of the corporate agenda at the international and national levels (Cohen, 1997). Liberal free-market strategies such as free trade agreements (for example, the General Agreement on Tariffs and Trade or GATT and the North American Free Trade Agreement or NAFTA) and Economic Processing Zones (EPZs) reduce fiscal and regulatory burdens on corporations, facilitate the movement of capital, and encourage export. In this context, transnational corporations have tremendous clout and the position of labour is weakened, as TNCs are less dependent on a nation's citizens as consumers and as workers (Cohen, 1997: 7). Bakkar (1996: 5) writes: "The increased structural power and mobility of transnational capital has created pressure on nation-states to be tax competitive and has sharply eroded national governments' power to tax owing to the spread of offshore tax havens." Nations now compete for corporations to set up shop in their backyards by reducing labour and environmental standards and promising cheap labour. In the North, this has meant a decline in jobs in the primary, manufacturing, and goods-producing sectors as jobs go South where structural adjustment policies (SAPs) based on free-market and deregulation principles are imposed (Connelly et al., 2000: 64-6).

Further still, arguments to justify restructuring are often linked to deficit reduction (Bakkar, 1996: 5; Cohen, 1997: 9-13). This argument contends that expensive social programs have pushed national economies into debt. However, this argument ignores the reality that the national debt in Canada, for example, is the result of a reduction in corporate taxes and escalating debt-servicing costs due to monetary policies, not expensive social programs. It assumes, unjustifiably, that people would be better and more efficiently served through the private delivery of services – via the private sector, the community, or the family (Connelly and MacDonald, 1996: 84; Luxton, 1998: 60). Neo-liberal advocates argue that these directions spell government surpluses, corporate profit, and individual wealth, and nations must follow suit or be left behind (Luxton, 2001: 4). Leach (1997: 36-7) points out that, at least in Western industrialized nations, economic conservatism has been accompanied by an emphasis on tradition and hierarchy – in contrast to the liberal ideas of equality and individualism – and an argument that advocates a return to the patriarchal family. Using the rhetoric of "family values," this New Right political philosophy maintains that society's ills are the result of the deterioration of the family, which leads to heavy reliance on state programs. The New Right philosophy insists that because public services are inefficient and expensive, and serve to undermine family and community ties, the responsibility and costs must be transferred from the welfare state to the patriarchal, self-reliant family. In this way, *both* the economy and the family are restructured.

The neo-liberal economic and political agenda marks a dramatic shift from the redistributive and collective direction of the

Keynesian welfare state to a preoccupation with cost-saving and the individual (Luxton, 2001: 3). Despite the claim that neo-liberal economic arrangements are more efficient, poverty, insecurity, and unemployment are the usual consequences. Cohen (1997: 5) writes: "Social and economic well-being for people is subordinate to the well-being of the corporate sector." Restructuring in Canada has been associated with the transformation of good jobs into bad jobs (Armstrong, 1996), or to put it another way, "occupational skidding" (Leach and Winson, 1995), marked by the growth of job insecurity and of non-standard, non-unionized work paying lower wages and providing no benefits, and by increased polarization of work hours. This trend is evident in Newfoundland fishing communities as well (see Neis, 2001; Neis et al., 2001).

Because women occupy a subordinate economic position due to caregiving responsibilities, which interrupt careers, and discrimination – demonstrated in the wage gap and occupational segregation into ghettoized jobs – these changes are having a disproportionate impact on them (see Leach and Winson, 1995). Women are doubly hurt – both as workers and as primary caregivers – by restructuring of health and social services (Connelly and MacDonald, 1996). Women have traditionally found good jobs in the public sector and rely more heavily than men on social services. Retrenchment of services does not mean that the need for such services disappears. Rather, it means the work gets redistributed to the domestic sphere (Connelly et al., 2000: 69). Armstrong (1996: 29-30) argues that restructuring in Canada is marked by a feminization of the labour force as more and more women enter the workforce, move into traditionally male areas, and experience increased similarity in employment with men. Yet, this does not necessarily mean real gains for women as men are increasingly doing work similar to that of women – non-standard work, lower-paying, insecure – and more women work due to economic imperative. Men's economic position has declined due to restructuring, rather than competition from women. In fact, as full-time jobs especially in manufacturing and primary industries disappear, men are successfully competing with women for traditionally female jobs – another example of the patriarchal dividend.

While the Keynesian welfare state has not been particularly pro-woman (Neysmith, 2000: 5), the shift towards the New Right has exacerbated women's unpaid domestic work. A solution to society's ills that relies on the patriarchal family makes women responsible for the reproduction of the family in ways that appear natural and thus devalued and unrecognized (Leach, 1997: 37). Even the more "generic" neo-liberal discourses make assumptions about what women and men do and what they should do. Like Keynesian economics, modernization approaches based on neo-liberal economics assume a male breadwinner (Bakkar, 1996: 6-7). Consequently, the rational economic man is the subject of much policy, and educational and work programs are aimed at men (Bakkar, 1996: 9). Local-based studies vary in specificities yet clearly demonstrate the gendered character of neo-liberal restructuring. For example, Ferguson (2001) demonstrates how Ireland's new economic growth and its integration into the global economy have both challenged and reinforced "traditional" hegemonic constructions of masculinity. Traditional constructs, shaped by the Catholic Church, emphasized rurality and hard work and the good family man as provider. Contemporary constructs challenge the Church, with increased weight placed on individualization. Yet, working hard and being the provider are reinforced, discouraging male participation in and assuming female responsibility for caregiving. This is perpetuated ideologically – reflected in men's long working hours – and through workplace and government policy. Munk-Madsen (1997) deconstructs the gender-neutral language used in quota regulations by the Norwegian fisheries management. She demonstrates how criteria for quota allocations privilege men and create barriers for women – the processes of which are very similar to the case of Newfoundland. At the same time as neo-liberal economic approaches ignore women, they simultaneously assume the existence and elasticity of their labour. Connelly (1998: 147) has argued that the neo-liberal agenda transfers the "costs and workload from the paid workforce to the unpaid household economy and onto the shoulders of women where it can not be seen" and relies on a gendered division of labour that forces women to "underwrite current policies with their low paid and unpaid labour."

There is increasingly a global imposition of both a gender ideology, which makes women primarily responsible for unpaid, low-paying, and reproductive work, and an economic ideology that ignores or devalues this work yet relies heavily on the gendered division of labour in which it is performed. Therefore, women in the North and South bear the brunt of restructuring as it is currently constructed. The male bias in restructuring exists around the world, in developing and developed regions, because women, their reproductive labour, their structural positions, and the conditions of their daily experience have been either excluded from consideration or taken for granted in policy- and decision-making processes.

## IS THERE A CRISIS OF MASCULINITY?

This book starts from a position of "crisis" and tries to account for the impact of and reaction to crisis. The notion of crisis presented here must be distinguished from its use in the "masculinity crisis" literature (for example, Bly, 1990) and the anti-feminist factions of the men's movement (Clatterbaugh, 1997). Whether this crisis is perceived to be rooted in the erosion of separate spheres, the successes of the women's movement, the separation of fathers from sons, or changes in the economy, it is assumed that the ideological and material bases of male power and advantage are crumbling. While the legitimacy of patriarchy is being questioned, a widespread breakdown has not happened (Brittan, 1989: 184). Rather, dominant configurations of masculinity seem to be changing form and adapting rather than dissipating (Clatterbaugh, 1997: 206; Connell, 1995: 211). Individual men may feel powerless; yet, practice is configured in ways that confer space for men to exercise power. Thus, "men's experience of powerlessness is *real* – the men actually feel it and certainly act on it – but it is not *true*, that is, it does not accurately describe their condition" (Kimmel, 1994: 137).

There are, however, local crises – like the current crisis plaguing the Newfoundland fishery – that may impact how masculinity is constructed and practised. For example, crises in production relations, including mass unemployment or reorganization of male-dominated industries, may create a number of obstacles for men to live their lives *as men* in culturally prescribed ways. Such crises in the world of work and the existence of few other meaningful roles may encourage men to question their identities and engage in negative behaviour (Cleaver, 2002: 3-4). When Davis (1993) argues that male fishers from the southwest coast of Newfoundland have become "feminized," she is not so much wrong as incomplete. Male fishers may *feel* less than men because they can no longer "be men" in culturally prescribed ways and they may compensate or react by engaging in negative behaviours such as abusing alcohol. However, this does not necessarily mean that they have become "women." In fact, Davis provides evidence to the contrary – men have adapted masculinity rather than abandoning it as a way to cope with their changed economic and cultural circumstances. Just as local crises impact how men behave and think, so, too, are the effects of crises mediated by local constructions of masculinity. In other words, men are able to deal with and respond to crises in historically and culturally specific, gendered ways.

*The Local and the Global*

Connell (1998: 15-16) and others (see Kimmel, 2001) argue that the current world gender order increasingly reflects a dominant neo-liberal agenda that privileges the male entrepreneur, transnational capital, and global markets. The result, he contends, is a hegemonic "transnational business masculinity" marked by a commitment to rationality, capitalist accumulation, and a lack of social responsibility, which may or may not conform to local gender orders. As such, responses to "transnational business masculinity" at the local level have been varied. Connell (1998: 18) and Kimmel (2001: 30-4) document local and global instances of opposition to male hegemony especially articulated through international agencies and networks. Other local responses have focused on protecting or reclaiming the "patriarchal dividend" through explicit masculine fundamentalism or through so-called neutral initiatives such as cuts in social spending in the name of global competitiveness, which inevitably hurt women because of their social position (Kimmel, 2001: 26-9; Connell, 1998: 17).

Despite this documented variation, Connell (1998: 19) then makes an unjustified leap, suggesting a shift in the research agenda from a focus on local and comparative work to the global arena. Connell's overemphasis on the global arena implies a sort of inevitability and linearity and runs the risk of essentialism. While the global context is increasingly important in any account of gender, I, like others (see Bakkar, 1996; Connelly et al., 2000; Pringle and Pease, 2001), maintain that while the global prescription is the same, the impacts of and responses to this prescription are constrained, contested, and mediated by specific political, cultural, and historical arrangements organized at the national and regional levels. Connelly et al. (2000: 78) write:

> all societies have been deeply and fundamentally affected by global capitalism, and for several centuries none have operated independently of the global economy. Quite evidently, globalisation has not meant that all societies have become the same, economically or culturally. Diverse local histories have emerged from particular interactions of the local and the global as people have accommodated, and resisted, the conditions they encountered and have pursued their daily activities in culturally meaningful ways.

A focus on the local – especially when that local area is rural and maritime – demands that attention is paid to how uneven development mediates the impacts of restructuring. Several characteristics of rural areas make them more vulnerable to restructuring – whether industrial, political, environmental, or social – than urban areas (Leach, 2000: 210-14; Leach and Winson, 1995: 349). Rural areas tend to be more dependent on single industries, especially extraction and manufacturing; they have fewer local job opportunities and farther commutes for new jobs; rural areas receive the "bad jobs" rather than "good jobs" associated with the service economy and are heavily dependent on public spending. Rurality also shapes available discourses, with implications for the construction of identity. Local people in Newfoundland use the maritime environment and rurality to define themselves in relation to what is "urban" and to make sense of change and form responses to it.

Like "malestream" approaches to development, early feminist analyses wrongly assumed universal effects of a global economy (Connelly et al., 2000: 90-8). Such grand theorizing ignores how local, historically specific contexts mediate global arrangements and their impacts, and denies agency in people's efforts to react. Although the global gender order favours men over women, not all women experience restructuring or global capitalism in the same ways around the world – nor do all men. Rather, women's and men's gendered experiences are, in part, mediated by a region's or nation's relationship to global capital – as well as the variegated social positions held by women and men, and the local meanings and possibilities attached to these positions. Neysmith (2000: 2) writes: "Women share a common responsibility to care for society's members who are defined as dependent; women differ in their social locations – and therefore the resources available to them – to carry out these socially assigned responsibilities." In other words, women manage caring work under different circumstances and therefore experience the impacts of restructuring in different ways. Similarly, while men generally may gain from the patriarchal dividend, this unearned advantage is mediated via other social locations and historically specific settings.

## COPING STRATEGIES

It is clear that the global economy and restructuring shape and get filtered through the local. People cope with the new arrangements, which, research in Canada and elsewhere suggests, make getting by more difficult for all but the elites. This new order requires families to take on even more responsibility for their own subsistence, using the ideas, practices, and resources available to them. Of course, relying on a combination of paid work, unpaid domestic and subsistence work, and the welfare state in order to survive is not new (Luxton, 1998: 57). Capitalism sets up an inherent contradiction by splitting the production of goods (and the costs) from the reproduction of people (and the costs) (ibid.,57-8; Luxton, 2001: 2). In other words, the economy depends on assiduous workers – not distracted by non-work-related concerns; yet they rely, in part, on unpaid domestic work to produce these workers on a daily and generational basis and to underwrite the costs of capitalism. This separation of production and reproduction means that unpaid work constrains opportunities to work for pay and that wages do not meet the costs of social reproduction on a daily basis, including "the provisioning of people to ensure they eat, sleep and get the other material and emotional sustenance they need to stay alive and keep going from day to day" or on a generational basis, including "the biological production, care and socialization of children" (Luxton, 1998: 58). The state and the family have tended to offset capitalism's shortcomings – the state through social welfare programs, including universal education and health care, and the family through unpaid domestic and subsistence work (ibid.; Luxton, 2001: 2).

Locally specific studies clearly demonstrate how families, households, and individuals work out coping strategies – ways of managing these shortcomings to make sure all the necessary productive and reproductive work gets done – by using the material and ideational resources available to them, including or perhaps especially those that configure practice by gender. Sexual divisions of labour and the gender ideologies that maintain them shape who does what work. For example, Skaptadottir and Proppe (2000) show how existing gender divisions in Icelandic fisheries organize how women and men cope with restructuring, and how gender makes some problems more relevant to men than women and vice versa. In their studies of the impact of restructuring and hard times on white, working-class families in Hamilton, Ontario, Luxton and Corman (2001) and Leach (1997) describe people coping by relying on the family – meaning women's unpaid work – and positively assessing their abilities to be self-reliant. The working-class male steelworkers and their wives explicitly endorsed conservative views about gender roles – assigning women to the domestic sphere and men to paid labour. Traditional gender ideology, women's real responsibilities for caregiving work, and structural inequalities and discrimination in the labour market shaped men's and women's possibilities for paid and unpaid work (Luxton and Corman, 2001: 33-4). In the context of job scarcity and insecurity and insufficient incomes, support for one job per family and the male breadwinner prerogative gained support among these families, which resulted in tensions between men and women in workplaces. Yet, there are occasions where people actively choose or are forced to contradict accepted practices and ideas. Despite ideological resistance, Luxton and Corman (2001) and Leach (1997: 40-1) found that material need compelled women who aspired to stay at home to work for pay and led to changes in the sexual division of labour in households – with men doing more. Fisheries-dependent households in Newfoundland have faced similar contradictions, as evidenced by the tensions experienced when women move into the harvesting sector, a traditionally male domain, as a coping strategy.

The coping strategies of the women and men in the studies by Leach and Winson (1995) and Leach (2000) were complicated by rurality. In these studies, the authors found plant closures in rural Ontario had differential impacts on local men and women who had worked in factories side by side prior to being let go. The local economy offered mostly "bad jobs" and acquiring better jobs required long commutes outside of the community. The latter was more likely an option for men than women due to child-care responsibilities (Leach and Winson, 1995: 357). The illusion of egalitarian gender relations while women worked in factory jobs alongside local men is shattered in the observation that men's and women's possibilities for new employment were differently shaped by the restructured rural economy and gender ideology. In their case study of restructuring in rural Newfoundland fisheries-dependent communities, Neis et al. (2001) demonstrate that women's and men's divergent experiences of displacement proved dangerous for women's health. Inadequate retraining and fewer job opportunities disadvantaged women. Similarly, MacDonald and Connelly (1990a) claim that household and subsistence work are related to the productive sphere in their study of a Nova Scotia community in the early 1980s. People responded to the restructured fishery not only as individuals or groups of workers but as members of households. Individuals made decisions about entering paid work based on the economic positions of other household members. For example, in this particular case study, fishers' wives were less likely to take up work at the plant because of fishers' increased incomes. On the other hand, low incomes for male plant workers were linked to women's entry into the plant for a second income (MacDonald and Connelly, 1990a: 144-5).

As neo-liberal policies compel households to do more subsistence and unpaid work to meet the needs of their members, not all members equally share this responsibility or have equal access to resources. In hard times, caregiving is both needed more and more difficult to give. Luxton (2001) argues in her study of people in the greater Toronto area that neo-liberal policies constrain what sorts of care people give and receive. Economic restructuring and government cutbacks mean that people's ability to provide care is limited by a lack of money just when the household has more unpaid work to do in the absence of the existence of satisfactory services. This causes stress and overload for household members, especially women, who are made responsible by gender ideology and "traditional" sexual divisions of labour. So while people are coping with deteriorating standards of living, the economic and interpersonal costs are great (Luxton, 2001: 7-9). Luxton and Corman (2001) also report stress in households, especially among women. In her analysis of the experiences of Nova Scotian male offshore fishers and their female partners, Binkley (1995: 89-90) finds that women – like the women in this study – often interpreted in a negative manner the increased male presence at home due to displacement. While some women enjoyed the opportunity presented since the cod moratorium of the early 1990s to spend more time with male partners, most women reported dissatisfaction, a felt loss of independence, and an increased workload due to heightened financial insecurity and the additional caring duties associated with having an unemployed fisher husband at home. The literature review appears to support the claim that care and emotional support are part of the patriarchal dividend men receive in hard times.

A body of research shows how structural unemployment may foster among men a sort of "radical pragmatism" to make ends meet (Connell, 1995: 97). Connell observes that the workplace and labour processes are not the only sites in the economic realm in which men construct masculine practices and identities. Rather, the conditions of capitalism and the "*labour market as a whole*" are sites of negotiation (Connell, 1995: 95). For male fishers in this study, this includes their participation, or lack thereof, in the formal and informal sectors, paid and unpaid labour, legal and illegal economic transactions. Morris's (1991) displaced British miners and Connell's (1995: ch. 4) unskilled and unemployed Australian men drew on personal connections and relied on informal and criminal activities to get by. Mac An Ghaill's (1994) work reveals how racial exclusion and discrimination, white authority, and the marginalization of black history and culture make it impossible for members of the Rasta Heads, a group of Afro-Caribbean young men in England, to adopt white working-class masculinities characterized by an investment in manual work and breadwinning (Mac An Ghaill, 1994: 191). In response to structural unemployment and declining manufacturing industries, which disproportionately affected the Rasta Heads, these men defined masculinity by toughness, Africanization, and non-conventional work (Mac An Ghaill, 1994:

186-9). Forrest argues that Cuba's changing position in the global economy via state promotion of tourism has produced the collusion of women and men in accepting racialized sexual stereotypes. An influx of Western tourists has created awareness of material deprivation among Cubans, while the legalization of the American dollar has created new forms of social inequality (Forrest, 2002: 97). In the context of scarcity and few alternatives, young locals use the Latin lover stereotype in the informal sex tourism industry (ibid., 2002: 92). This body of research describing coping strategies highlights the significance of both structure and agency. Changes in the local economic conditions impact gender configurations of practice. At the same time, culturally specific gender practice and ideas provide the tools with which men and women are able to respond.

*Consciousness, Accommodation, and Resistance*

Available material resources and discourses shape people's daily lives, identity-formation, and how they make sense of the world. A focus on the local ideational and material configurations of gender and on how these are mediated through the household helps to keep agency in the analysis and avoid economic determinism. Women and men make decisions about how to cope with unemployment and economic vulnerability not as isolated individuals but in social contexts and relations, and as part of social groups, and possibilities in both the formal and informal sectors of the economy are enabled and constrained by accepted ideas and practice. Coping strategies are also sites in which to create meaning and a sense of self or identity, and they can become sites of resistance and accommodation. Coping strategies are not simply about meeting physical needs; rather, they are about both access to and the distribution and control of resources *and* ideas. As men find it increasingly difficult to conform to the breadwinner role or other hegemonic versions of masculinity, how they assess their situations is linked to possibilities for change. In other words, consciousness is affected by material arrangements and available discourses and possibilities in relation to gender, age, race, and class.

Socialist accounts of capitalist work environments (for example, Clatterbaugh, 1997; Collinson, 1992; Tolsen, 1977; Willis, 1977) reveal contradictions for working-class men. Willis (1977: 125) argues that the exploitation of labour for profit and the commodification and deskilling of labour make workers aware of their subordinate economic position and expendability, and consequently they can develop "radical insight" to see through dominant ideologies, institutions, and structures in wider society. Conversely, Collinson

(1992: 79-80) says that the position of the working class is not always conducive to the construction of oppositional insights. Rather, because insecurity of the self is a major motivation for group organization, membership is both oppositional and conformist. These conditions, combined with the separation between worker and the final product in the capitalist workplace, encourage men to take an instrumental approach to work rather than investing their sense of self in work (Willis, 1977: 99-102).

This ambiguity sets up a range of possible reactions to the effects of restructuring. According to Yarrow (1992: 55-6), throughout the twentieth century worker consciousness among Appalachian coal miners oscillated from accommodating to challenging. In periods of layoffs and mine closures, Appalachian miners replaced their investment in a masculinity that focused on toughness and resistance with one marked by compliancy in order to secure work and maintain dignity. Marxists who have investigated the fisheries have tended to see the contradictory class position of inshore fishers – as (semi-proletarian) dependent commodity producers – as space for resistance. Yet, MacDonald and Connelly (1990b: 155) argue that the notion of class, based on a single person's relationship to the means of production, is problematic. Just as women's class position is contradictory and shaped by their "direct and indirect relations to the means of production – their own productive and reproductive labour as well as that of their husbands," partners, or family members, so, too, the class position of men is contradictory. MacDonald and Connelly insist that the household is central to understanding why people behave in certain ways and not others. In other words, individual work histories, the work patterns and strategies of household members, and potential conflicts between members influence class identification, action, and inaction.

At the same time, and like Willis's working-class lads, Yarrow's (1992: 55-7) miners invested in a masculine identity that values mental over manual work, which served to separate Appalachian miners from management and owners, but also women, potential supporters in the class struggle, and thus tended to reproduce the status quo rather than challenge existing structural organization. The tendency for working-class men to underscore sexual difference – through the breadwinner role, sexual divisions of labour, or misogynist talk and practice – has been noted by others (Collinson, 1992: ch. 3; Tolson, 1977: 70-1) and serves as a weak basis for solidarity. Similarly, Connell considers the likelihood of successful collective resistance among unskilled and unemployed Australian men – with few, if any, opportunities for breadwinning – as slim because of their investment in compulsory heterosexuality. Connell (1995: 111) interprets this behaviour as a way to claim power through overt gender display where "there are no real resources for power."

Potential for resistance is curbed when age or race and ethnicity are used to define exclusive group membership. Collinson's shop-floor workers (1992: ch. 6) engaged in self-differentiating strategies based on age. Yet, these strategies weakened the power of the group in dealing with management by creating individualism and divisions among the workers. Older workers differentiated themselves from younger workers in ways that served to impose self-discipline and compliance to authority, which impeded collectivity. A sense of difference among men along the line of age may reflect their "cohort-specific, gendered social lives" (Thompson, 1994: 7). Soloman and Szwabo (1994: 42-3) argue that older men respond to new circumstances in ways that are consistent, at least subjectively, with their life histories. In Luxton and Corman's work (2001), the families of white working-class steelworkers in Hamilton, Ontario, responded to increased job insecurity by devoting increased time and effort to work – for less compensation – and to the family – for free. Male and female respondents viewed themselves as "ordinary" people bearing the brunt of economic restructuring and failed to recognize their ethnic privilege and, in the case of men, the patriarchal dividend (Luxton and Corman, 2001: 6). Fine, Weis, Addelston, and Marusza illustrate how poor and working-class white boys and men in the United States make sense of the effects of a declining manufacturing sector by ignoring their racial and patriarchal privileges and blaming specific groups of "others" – women and African Americans – for their deteriorating economic positions and for "stealing" their jobs through affirmative action policies (Fine et al., 1997: 54).

Another obstacle to collective resistance is embedded in global capitalism itself – the mobility of capital and workers makes identifying the source of workers' exploitation difficult. Margold (1994: 27) uses this explanation in her analysis of the migratory workers from the rural Ilocos region in the Philippines who interpreted their work experiences abroad as personal failures. Margold (1994:19) argues that local masculinities were "partially disassembled" when transnational corporation-led global restructuring changed the opportunities to live out ideal constructions of masculinity. While Ilocos men had a history of migration for labour, contemporary migration to the rich Arab Gulf shifted from being a choice to an economic necessity, from being a rewarding experience – in terms of the quality of work conditions and relations – to being an alienating, dehumanizing, and racialized experience (ibid., 34).

When people can identify a visible "enemy" (ibid., 28), have access to an alternative analysis of their situation, and are able to connect their problems to society, they are more likely to engage in political social action. The Canadian literature (see Bakkar, 1996; Brodie, 1996; Cohen, 1997; Leach, 1997; Luxton and Corman, 2001) suggests that while most people recognize and are critical of current New Right political and economic directions, they lack the tools to identify the causes. It is not surprising, then, that job insecurity is accompanied by a "survival orientation" that focuses on self-interest and family-centred solutions, and decreased support for equity policies – indicating the success of new corporate and familial hegemonies. Simultaneously, there are spaces of contradiction that provide openings for agency and opposition. For example, Leach (1997: 43) observes that in efforts to protect their jobs, male steelworkers responded to job insecurity with weak union militancy but were not willing to sacrifice seniority, and that men and women accepted that women should not compete with men for jobs but recognized the material need that women must work. People's behaviours, then, do not always match what they believe, especially when they perceive explicit resistance to be too risky. In their work on Icelandic fishing communities, Skaptadottir and Proppe (2000) found that men fishers participated in the reorganized fishery, despite ideological opposition, for fear of exclusion. They did, however, like the young men in Willis's study and like some of the fishers in the present study, engage in behaviours that can be read as symbolic resistance.

## APPLYING THE LESSONS LEARNED

Based on the literature reviewed in this chapter, it is clear that the processes and effects of restructuring are gendered, as are the responses to those effects. In the remainder of this book I apply the GAD approach (see Connelly et al., 2000) to make sense of male fishers' coping strategies and responses to the current direction of restructuring of the Newfoundland fishery. The GAD approach to restructuring opens space to investigate the lives of men through a gender perspective. It is best understood as feminist political economy that asks questions about the material and ideological bases of inequality between women and men. I examine the condition (material state) and position (relative status) of women and men in fishing communities and ask who has access to and control over resources, who does what work – productive and reproductive – and how is this work valued. I also consider how structures and discourses define and maintain women's general subordination and the patriarchal dividend for men. However, the social positions of individual women and men vary, which means that men experience patriarchy differently, as do women, and are shaped by intersections with other configurations of practice (such as class and age) as well as by local, historical, and cultural contexts. How male fishers make sense of these processes shapes their consciousness, identities, and thus the possibilities for action.

The men and women studied here responded to crisis and restructuring in ways that reproduced patriarchal arrangements and other inequalities. Male fishers' coping strategies and responses to restructuring initiatives reflect a desire for continuity in a tumultuous time, as well as the availability of particular tools – both material and ideational – to resist and accommodate change. Some strategies reasserted the "traditional male fisher" model found in a "looking to the past" framework. This framework enabled men to formulate both alternative and accommodating assessments of the fisheries crisis and the state's responses to it, and is marked by a realization that fishing offers rewards unavailable in other manual jobs, and, thus, was worth clinging to. Other strategies more neatly fit the business model of the "modern male fisher" embedded in state discourses, policies, and initiatives. This model assumes that the most efficient fishing industry is organized around the male entrepreneur discrete from the family. The contradictory class position of inshore fishers simultaneously constrains and facilitates their acceptance of such a model, as well as their consciousness and collective sense of identity.

Despite the gender-neutral language of state adjustment and restructuring, the pre-existing positions of and notions about women and men facilitate patriarchal consequences. The current criteria used to determine legitimate harvesters and the methods of distributing access to resources via individual quotas are serving to reinforce a history of male privilege and to solidify corporate interests. This study reveals that men fishers have been engaged in these processes with varying degrees of accommodation and resistance. While resistance to the corporate model has been strong, at least on an individual level, there has been little, if any, challenge made to the patriarchal dividend they gain by virtue of being men. Women, on the other hand, have been compelled to return to the home because of the gendered distribution of access to resources and their unequal status in the fishing industry, inadequate adjustment measures, cuts in social spending, and exclusionary changes in employment insurance. In a context where there are few opportunities for women to construct culturally "appropriate" feminine identities and to enjoy real material independence, Newfoundland women have actively responded by reconfirming responsibilities for and psychological investment in domestic and caring work. This reconfirmation explains, in part, women's reluctance to treat men's increased presence at home (as a result of imposed moratoriums and new regulations) as an opportunity to move towards a more equal distribution of housework. Economic restructuring, then, is mediated through the household or family, reproducing unequal gendered divisions of labour, increasing burdens for women to shoulder, and reinforcing oppressive gender ideologies.

# The Place, the People 2

## A Brief History of the Newfoundland Fisheries

### THE SIGNIFICANCE OF THE PAST

> Newfoundland would have been paradise today if we hadn't go into Canada. Canada took it all, everything we had, took it all . . . we had the best fishery in the world. . . . But after Confederation, and Canada and they took over, our fishery was declining ever since. (NGP-3, retired male fisher, age 71)

Like the fisher quoted above, the men and women presented in this book used "the past" as the major referent against which the current fisheries crisis and economic restructuring are assessed. Some described the past as idyllic; others, dysphoric. At times, the past was conceptualized as distant; at other times, more recent. No matter which version was adopted, both independently and simultaneously, each provided fisheries people with tools to help formulate material and ideological responses to the current crisis. Indeed, the notion and reality of crises are entrenched in the history of the Newfoundland fishery (Fishery Research Group, 1986: 5). Male fishers in particular used the history of crises and the "traditional" inshore fishery to lay claim to certain rights of access, justify divisions of labour, and create a masculine identity.

### THE PRE-CONFEDERATION YEARS

The forefathers and foremothers of contemporary fishers of European descent settled along the Newfoundland coastline in the nineteenth century, transforming the largely migratory fishery[1] into a decentralized, household-based operation (Hutchings and Myers, 1995: 50; Neis, 1993: 190). Favourable weather conditions and the

"Relics from the past" available for camera-happy tourists visiting Bonavista (Photos by Nicole Gerarda Power).

migratory patterns of cod made summer and autumn optimal for fishing. Fishing households dealt with the intense fishing season by dividing work according to sex (Porter, 1993: ch. 3). Using labour-intensive technologies such as hook-and-line gear and later cod traps, and fishing from small open boats, men harvested cod for subsistence and barter. When not fishing or preparing to fish, men hunted, trapped, cut wood for fuel, and prepared the land for subsistence gardens. In many fishing communities, women participated in the production[2] of lightly salted, sun-dried codfish, in addition to tending gardens, raising cows and other animals for food, picking berries, caring for children, and performing domestic work.

Survival in the settler colony required the successful integration of productive and reproductive work in part because of the exploitative nature of the mercantile relations between fishing households and capitalist merchants who purchased salted fish (Ferguson, 1996: 31-3; Neis, 1993: 190-1). Supplying large quantities of salted fish did not ensure survival of the household, especially if merchants or their employed cullers deemed the supply to be of a low grade or quality – which fetched a low price – or if households had outstanding debts for supplies bought throughout the year on credit in local stores. Because merchants owned local stores and, thus, decided the price of local goods and determined markets,[3] households rarely received cash or credit. Simultaneously, merchants encouraged a patriarchal family form by refusing to deal with women in the exchange of fish and by providing credit to women for processing work only through their husbands, barring exceptional circumstances.

While merchants largely controlled what happened to fish out of water, fishers regulated the health of local fish stocks and resources through informal community-based management (Sinclair, 1988b: 157). Until the twentieth century the involvement of fisheries science in decision and policy matters relating to Newfoundland's fishery was limited (Ommer, 1998: 14-15). When the Newfoundland government used scientific knowledge to inform fisheries-related policies, it was done so in conjunction with local fishers' observations. Indeed, the Newfoundland government established the Fisheries Commission in 1889 in response to concerns voiced by both fishers and European scientists regarding the ecological health of fish stocks. Fisheries science gained a more secure place in government between 1900 and the 1930s with the recruitment of outside experts to facilitate the management and modernization of the Newfoundland fishery. In 1930, the Newfoundland Fisheries Research Commission was created to examine fish biology and processing methods (ibid., 18).

## CONFEDERATION, MODERNIZATION, AND CORPORATE OWNERSHIP, 1949-1969

In 1949 Newfoundland's fisheries were subjected to increased international regulation with the establishment of the International Commission for the Northwest Atlantic Fisheries (ICNAF) and state intervention through Confederation with Canada (ibid., 20-1; Sinclair, 1988b: 160). During the early years of the post-Confederation period, the provincial and federal governments, backed by corporate support, implemented policies and development schemes that encouraged the industrialization and rationalization of the fishery (Antler and Faris, 1979; Ferguson, 1996: ch. 7). Newfoundland's decentralized, household-based production of salt fish was gradually replaced by a vertically integrated processing industry characterized by factory-based mass production of semi-processed blocks of fresh and frozen fillets[4] (Sinclair, 1987: ch. 4). This shift was undoubtedly facilitated by the changes in market demand after the introduction of quick freezing technologies in the 1930s and again after World War II (see Wright, 2001: 20-2). The dominant view holds that the decline of the salt fishery was the result of a degeneration of the quality of curing skills and the introduction of cod traps that increased landings (Ferguson, 1996: 263-5). However, many authors have convincingly linked this decline to purposeful directives on the part of the state, which intended to transform members of "backward" inshore fishery-dependent households into participants in a "modern" economy (see Antler and Faris, 1979; Ferguson, 1996: ch. 7; Wright, 1995).

During this period, the provincial and federal governments financed projects that aimed to modernize both the processing and harvesting sectors (Sinclair, 1987: ch. 4). Both levels of government promoted technological development by introducing longliners to the inshore sector, offering loans and subsidies for "modern" fishing technologies, and supporting the development of industrial ownership (Ommer, 1998: 20-2; Wright, 1995: 129-30). Modern processing was facilitated with the construction of frozen-fish processing plants around the province, and labour for these plants was supplied via the provincial government's resettlement program, which relocated people from isolated outports[5] scattered around the island's coastline to "fishery growth centres" (Andersen, 1972: 120). Resettlement, which disrupted traditional crew recruitment patterns and removed people from ocean and land resources traditionally used for subsistence, and the construction of community stages, which made the household-based production of salt fish impossible, supported the fresh and frozen-fish industry and encouraged the withdrawal of merchants (Neis, 1993: 201; Nemec, 1972: 32; Wright, 1995: 130). The merchant credit system was replaced by cash payments, further rupturing household strategies that relied on an integration of both informal and formal economic activity and occupational pluralism (Neis, 1993: 201; Ommer, 1998: 19). The extension of the post-war Keynesian welfare state also had the effect of disrupting the household-based salt fishery. Cash inputs, derived from such transfer payments as family allowances, unemployment insurance, and old age pensions, altered the informal barter and subsistence economy (McCay, 1988: 112; Neis, 1993: 195; Ommer, 1998: 19). Compulsory education for children and job training aimed at men, both of which subsidized the reproduction of the labour force, and the dissemination of a gender ideology, which increasingly defined women as home-based wives, mothers, and consumers, normalized the heterosexual patriarchal nuclear family (McCay, 1988: 112-13; Wright, 1995: 133, 141-2).

Finally, many fishers and members of their households were encouraged to leave the salt fishery and take up employment in new land-based industrial jobs during this period due to declining fish prices and incomes, and legal obstacles that limited alternative work opportunities for fishers and impeded unionization (Neis, 1993: 201-2; Ommer, 1998: 19). Encouraged by state financial support and the extension of unemployment insurance (Ommer, 1998: 22), those "independent" inshore fishers who remained in the industry responded to these obstacles by intensifying exploitation. Fishers invested in larger and faster boats and procured new gear to increase harvesting efficiency (Neis, 1993: 201; Andersen and Wadel, 1972b: 156). Other fishers took up work in the deep-sea harvesting sector (Sinclair, 1988b: 159-60). While their incomes may have been better than those of inshore fishers, the status of offshore crew members as risk-sharers meant they lacked any formal minimum-wage and labour protection, endured dangerous working conditions, and spent long periods at sea, which prevented them from participating in traditional subsistence work.

The inshore fishery, then, took place alongside a corporate-owned offshore Canadian trawler fishery, as well as European factory freezer trawler fleets (FFTs), which fished offshore waters with tremendous harvesting capacity (Hutchings and Myers, 1995: 58, 64; Neis, 1991: 147; Ommer, 1998: 22). While Canada did not invest substantially in FFTs until the late 1970s (Hutchings and Myers,

1995: 64), freezing and filleting technologies developed for FFTs were incorporated into the land-based factories to produce fish blocks for the United States market (Neis, 1991: 161). Reliance on the mass production of a single commodity for a single market, however, proved problematic by the 1960s (Neis, 1991: 161-3; Sinclair, 1988b: 159). While the production of fish blocks does not require consistent size or quality of raw material, it does demand a dependable supply of fish. Barring the fishery on the south coast, the seasonal nature of Newfoundland's fishery, running from June to October, and dependency on weather and cod migratory patterns made this demand impossible to meet. In an attempt to overcome the problem of procuring dependable and year-round supplies of raw material, companies began to rely on both inshore fishers and offshore corporate-owned trawlers for labour and supplies of fish. The mass production of block product did not require complex mechanization, high levels of supervision, or job fragmentation, and was characterized by wasteful machines and discard practices. Fishing firms maintained direct control of productivity by relying "on the pressures of a backlog of work and limits on the freedom to strike" (Neis, 1991: 163). Companies extended their control over workers and inhibited the organization of fisheries workers through multi-plant ownership, nepotistic and gendered hiring practices, ownership of other community-based businesses, and the construction of plants in areas with a reserve army of flexible but unskilled labour, that is, women workers (ibid.).

The wasteful relationship companies had to resources[6] and ICNAF's ineffective regulation of FFTs, as well as local and foreign fishing, had contributed to stock declines by the mid-1960s (Neis, 1991: 160; Sinclair, 1988b: 160). As a result, from 1964 until the 1970s the number of inshore fishers declined, as did their catches of cod and their incomes, while the offshore sector experienced declining landings of redfish and flatfish (Hutchings and Myers, 1995: 71; Sinclair, 1988b: 159, 161). Government responded to declining landings, foreign over-fishing, and demands from inshore fishers by increasing state management as well as resources for fisheries science, and by instituting a 12-mile management zone in 1969 (Ommer, 1998: 21-2). The regulatory regime in the 1960s largely consisted of limited fishing seasons, minimum fish sizes, and restricted types of harvesting gear and gear areas to control exploitation. In addition, licence schemes were introduced in some fisheries to limit access. Nonetheless, rules concerning access to the fisheries resources and harvesting gears remained largely in the hands of local fishers at this time – there were few formal restrictions on catch levels or fishing capacity, on who could hold licences, or on the number of licences in effect (Department of Fisheries and Oceans, 2001: 8; Sinclair, 1988b: 157).

### THE ERA OF FISHERIES SCIENCE, STATE REGULATION, AND RESTRUCTURING, 1970s AND 1980s

Stocks continued to decline into the 1970s and 1980s (Hutchings and Myers, 1995: 72-3). Canada responded in the 1970s by extending limited-entry licensing to regulate access to most Atlantic fisheries. In addition, the rate of exploitation was managed through the regulation of harvesting technologies such as vessel replacement restrictions and the introduction of quota management in the form of total allowable catches (TACs) for most species of commercial value (Department of Fisheries and Oceans, 2001: 8-10; Sinclair, 1988b: 163-73). During this period fisheries science became entrenched in national management with the establishment of the Canadian Atlantic Fisheries Scientific Advisory Committee (CAFSAC) and the Department of Fisheries and Oceans (DFO) (Finlayson, 1994: 6-7). The creation of a 200-mile exclusive economic management zone in 1977 and national allocations for all major international groundfish stocks were introduced by ICNAF (later replaced by the Northwest Atlantic Fisheries Organization or NAFO) (Department of Fisheries and Oceans, 2001: 8).

At the same time, processing companies faced problems of supply and demand (Neis, 1991; 165-7; Sinclair, 1987: ch. 8). By the early 1980s, the major processing companies in Atlantic Canada faced bankruptcy. Plants encountered declining landings of fish and shrinking profits. Alternative protein products had replaced some of the market demand for standardized frozen-fish products. In addition, processing companies in Newfoundland lagged behind companies in other fishing countries in terms of the acquisition of new technologies for standardized production (Neis, 1991: 165). Nonetheless, increased state regulation of the fishery, the institutionalization of fisheries science, which provided optimistic predictions about fish stocks, and the establishment of the 200-mile limit, which expanded national exploitation opportunities, encouraged processing companies to tackle these problems by discontinuing mass production and restructuring the industry (Neis, 1991: 148-9).

The 1982 Task Force on the Atlantic Fisheries supported this direction. The authors of this report assumed that fisheries scientists had correctly predicted that the stocks were healthy (Finlayson, 1994: 9) and the introduction of automation and new expensive technologies would reduce processing costs (Fishery Research Group, 1986: 12). Following this report, most of the assets of the deep-sea companies in Newfoundland were consolidated into a single company, Fishery Products International (FPI), owned by the provincial and federal governments and the Bank of Nova Scotia – later reprivatized – and the state supported the rationalization of the offshore groundfishery with the introduction of enterprise allocations (Department of Fisheries and Oceans, 2001: 9).

Restructuring entailed the diversification of production to other species, the creation of new and specialized products, the introduction of new, more highly automated technologies to eliminate waste, and the reorganization of labour to deal with increased concerns about quality, all of which brought additional costs. Management responded to the problem of increased costs and quality control by drawing on local reserves of women as low-paid fish-processing workers and by introducing incentive systems that rewarded bonus pay for speed and quantity of output, as well as yield from raw material (Fishery Research Group, 1986: 13-19; Neis, 1991: 164-6; Rowe, 1991: 15, 21-2).

In addition, this renewed optimism in the fishery gave rise to the construction of more fish-processing factories. Changes in the production of fish and the construction of new plants opened up employment opportunities, especially in the form of irregular work (part-time, casual, or seasonal jobs) in processing, and encouraged more people to enter the harvesting sector in the 1970s and 1980s (Fishery Research Group, 1986: 70-1). Between 50 and 60 per cent of Newfoundland's processing jobs, particularly irregular jobs, were filled by women, and women made modest gains in accessing jobs in the harvesting sector, going from 8 per cent of fish harvesters in 1981 to 12 per cent by the early 1990s (Rowe, 1991: 18; Williams, 1996: 1, 23; Women in Resource Development Committee, 2003). Expanded employment opportunities and community-based labour shortages strengthened the position of plant workers during this time. Restructuring and unionization of fishers and plant workers also strengthened the inshore fishery in the 1970s (Neis, 1991: 164). Restructuring entailed an increased reliance on the decentralized inshore fisheries, which used technologies that were less damaging to fish than those used in the offshore fishery (Fishery Research Group, 1986: 14; Neis, 1991: 164). At the same time, federal government conservation initiatives thwarted expansion of the offshore fishery, at least temporarily (Neis, 1993: 202). The inshore fishery had always employed more workers than the offshore fishery and more communities depended on the former (Department of Fisheries and Oceans, 2003c).

Science-informed state regulation of the fishery, the state's bailout of processing firms, and the subsequent restructuring provided temporary stability in Newfoundland's fishing industry. However, inshore fishers and plant owners once again experienced declining landings of cod. By 1983 inshore fishers contradicted fisheries scientists' estimates of the health of stocks based on their observations of changes in fish size and their experiences of decreased landings, despite increased effort, time, and financial investment[7] (Neis, 1992: 165; Neis, 1993: 202-3). The results of a 1986 independent study commissioned by the Newfoundland Inshore Fishermen's Association stood in contrast to reports of high offshore catches of northern cod during the 1980s, no doubt reflecting the mobility of gear used in this fishery (Hutchings and Myers, 1995: 76; Neis, 1992: 157), and to the stock estimates produced by the Canadian government (Alcock, 2000). Inshore fishers' concerns were hardly given serious attention. Fisheries scientists responsible for stock assessment at the DFO regarded the ecological knowledge of fishers and data from the inshore as, at best, incompatible with their methodologies or, worse, unreliable and anecdotal. The scientists preferred to use catch and effort data from offshore research and commercial vessels. This neglect enabled consistent overestimations of the biomass of fish stocks and thus the allocation of excessive TACs (Finlayson, 1994: 104-6; Neis, 1992: 162).

## FISHERIES CRISIS, MORATORIUMS, AND THE STATE'S NEO-LIBERAL RESPONSE

Inshore fishers' concerns regarding the health of cod stocks were later echoed in the 1990 *Independent Review of the State of the Northern Cod Stock* (Harris, 1990), which found the methodologies and models used by DFO scientists to be flawed, underestimating fishing mortalities, and thus contributing to over-fishing and declining landings in the Atlantic fishery (Finlayson, 1994: 9). This report recommended a decrease in total allowable catches of northern cod for the waters off Newfoundland and Labrador (Department of Fisheries and Oceans, 2001: 11). In response to declining landings, the processing companies tied up their offshore corporate fleets and, with the assistance of the state, restructured the industry on a global scale (Alcock, 2000). The state responded with a number of downsizing efforts and adjustment initiatives. In 1992 the federal government made an extraordinary decision to place a moratorium on northern cod in areas 2J3KL[8] in response to declining stocks, and by 1995 moratoriums had been extended to all groundfish stocks around Newfoundland except redfish (Sinclair, 1999: 324).

Although moratoriums affected the east coast provinces, Newfoundland and Labrador was hit hardest due to its heavy reliance on the groundfishery, a weak economy, and an export-driven fishing industry (Alcock, 2000; Human Resources Development Canada, 1998: 75). Upwards of 30,000 fisheries workers in Newfoundland, or 12 per cent of the province's total labour force, were directly affected (Sinclair, 1999: 325). Alcock (2000) has effectively argued that the domestic political history of Canada is linked to the adoption of particularly detrimental fisheries policies that favoured the offshore and vertically integrated processing sectors. A number of factors have contributed to the failure of government to respond effectively to the crisis, including the federal system that splits authority for fisheries policy-making between federal and provincial[9] governments, the fact that the small-scale inshore sector accounted for less than one-third of the cod landed, fishers' heavy reliance on unemployment insurance, and their political fragmentation.

Originally, the moratorium on cod was predicted to last for two years, with eligible individuals receiving income replacement and alternative employment opportunities or retiring from the fishery as part of the Northern Cod Adjustment and Recovery Package (NCARP) (Fish, Food and Allied Workers, 1992: 6-7). DFO decided eligibility by determining dependence on northern cod based on previous income from work in the cod fishery. By 1993, however, the cod stocks continued to be in ecological crisis and the moratorium was extended to other areas of the province and the Atlantic region for an indefinite period. In 1994, following the recommendations of the Task Force on Incomes and Adjustment in the Atlantic Fishery, the state replaced NCARP with The Atlantic Groundfish Strategy (TAGS), the purpose of which was to direct recipients into alternative careers, to retire people from the fishery, and to provide temporary income support[10] (Human Resources Development Canada, 1998: ch. 2; Fish, Food and Allied Workers, 1994b: 3, 7; 1994a: 6). At the same time, however, adjustment was made more difficult. Funding for post-secondary education was slashed, private, more expensive schools replaced public colleges, and changes to the unemployment insurance program reduced access to benefits and training by means of tighter eligibility criteria (Fish, Food and Allied Workers, 2003; Neis et al., 2001: 19; Skipton, 1997: 18).

The direction of fisheries policies reflects a more general commitment to the downsizing of the bureaucratic state and to the free market. Neis and Williams (1997: 51) argue that state assistance may appear to contradict the neo-liberal agenda but that in fact it does not. Strict eligibility requirements justified by the assumption that fiscal restraint is necessary disqualified many of those who were affected. The intended purpose of such monies, after all, is to remove people from the fishery, and issues of scale and fisheries mismanagement are ignored. State initiatives did manage to remove 871 groundfish licences through the Northern Cod Adjustment and Recovery Program's licence retirement and early retirement initiatives and 545 groundfish licences through The Atlantic Groundfish Strategy's retirement schemes (Department of Fisheries and Oceans, 2001: 11). In 1998 the Canadian Fisheries Adjustment and Restructuring Program was introduced as the last attempt to reduce the number of groundfish licence holders through government-assisted retirement (Department of Fisheries and Oceans, 2001: 11-12). There has been a steady decline in the numbers of groundfish licences issued since 1991. In 1991 there were 9,447 groundfish licences in Newfoundland and 17,200 in all of Atlantic Canada, including Newfoundland. By 2000 these numbers had shrunk to 5,028 groundfish licences in Newfoundland and 10,783 in Atlantic Canada (Department of Fisheries and Oceans, 2003).

Such responses are part of the state's larger vision to restructure and eventually privatize the Newfoundland and Labrador fishing industry. The dominant view adopted by the state is that economic restructuring, informed by a neo-liberal agenda, is the solution to declining stocks and overcapacity. Adjustment initiatives accompanying moratoriums have thus focused on reducing debt and social spending, expanding export production, increasing regulation of access and entry, and encouraging withdrawal from the fishery with financial assistance. Implicitly and explicitly, the state has adopted the tragedy of the commons thesis:

> It has been recognized for almost 50 years that the heart of the problem of managing capture fisheries lies in the "common property" or "common pool" nature of wild fisheries. In pure "common pool" fisheries, access is open, property rights to the resources are virtually non-existent, and therefore fishers have no incentive to conserve the resource. (Department of Fisheries and Oceans, 2002: 5)

This view was used to justify increased state intervention in the management of the fisheries beginning in the 1970s. Mediated through neo-liberal ideology, it has more recently been used to argue the need for increased privatization of the ownership and management of fisheries resources. While the fisheries of Canada technically belong to its people and the constitutional authority for

Canadian fisheries remains the federal government (Department of Fisheries and Oceans, 1996; 2001: 42), a focus on debt reduction and international competitiveness lend support for the retrenchment of government services and the state's adoption of business models. The result has been the implementation of enterprise allocations and individual quotas, some of which are transferable, in some sectors and fisheries and the support of professionalization efforts that limit access, evidenced in the 1995 Montreal Round Table, an industry-government conference (Department of Fisheries and Oceans, 2001: 13). The establishment in 1993 of the Fisheries Resource Conservation Council (FRCC), which reviews stock assessments prepared by DFO and recommends TACs and other conservation measures, helped solidify this direction by emphasizing "co-management" and "shared stewardship" (Department of Fisheries and Oceans, 2001: 12).

## ECONOMIC AND POPULATION TRENDS POST-MORATORIUM

In many ways, Newfoundland and Labrador's economy looks a lot like that of a developing country – it is heavily dependent on resource extraction and international export, there are high levels of poverty and unemployment, and there is a disproportionate reliance on seasonal and low-paying jobs (Fishery Research Group, 1986: 116). Reflecting this situation, Newfoundland and Labrador's economy relies heavily on federal and provincial governments for direct and indirect employment and for income transfer payments (Human Resources Development Canada, 1998: 75). Between 1992 and 2002, Newfoundland and Labrador's unemployment rate peaked in 1993 at 20.4 per cent and declined to a low of 16.1 per cent in 2001, increasing to 16.9 per cent in 2002 (see Figure 1) (Economics and Statistics Branch, 2003: 7; Government of Newfoundland and Labrador, 2003). This is in stark contrast to Canada's unemployment rate in 2002 of 7.7 per cent (Newfoundland and Labrador Statistics Agency, 2003). In 2002, Canada enjoyed "its strongest job gains in 15 years," yet Newfoundland and Labrador was the only province not to experience increased employment (Statistics Canada, 2003).

Employment levels in Newfoundland and Labrador's fishing industry remain well below levels documented prior to 1992. Just prior to the initial moratorium, the industry employed over 20,000 people in 1991; in 2002, the industry counted about 16,200 employees (Economics and Statistics Branch, 2003: 19). Processing has been especially affected by reduced quotas, increased automation, and the production of a less labour-intensive product in crab processing (Department of Fisheries and Aquaculture, 2001: 1-2). In

Figure 1: Unemployment Rates, Newfoundland and Labrador

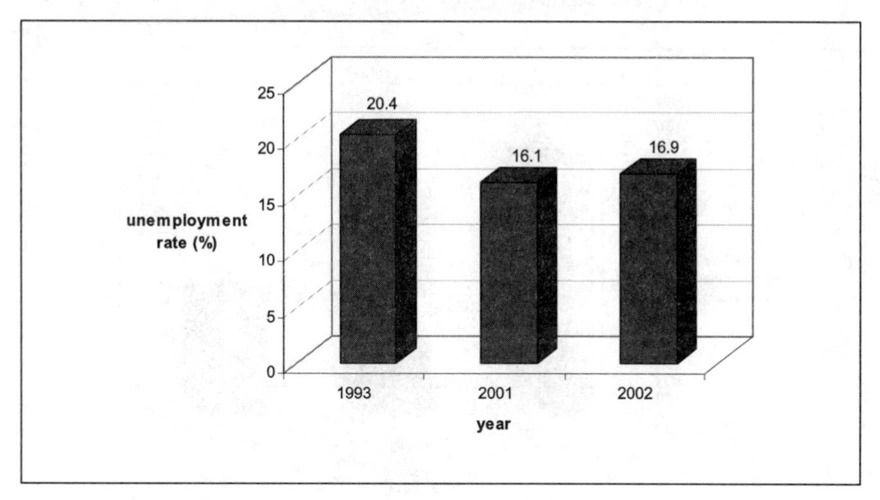


Figure 2: Number of Processing Plants, Newfoundland and Labrador

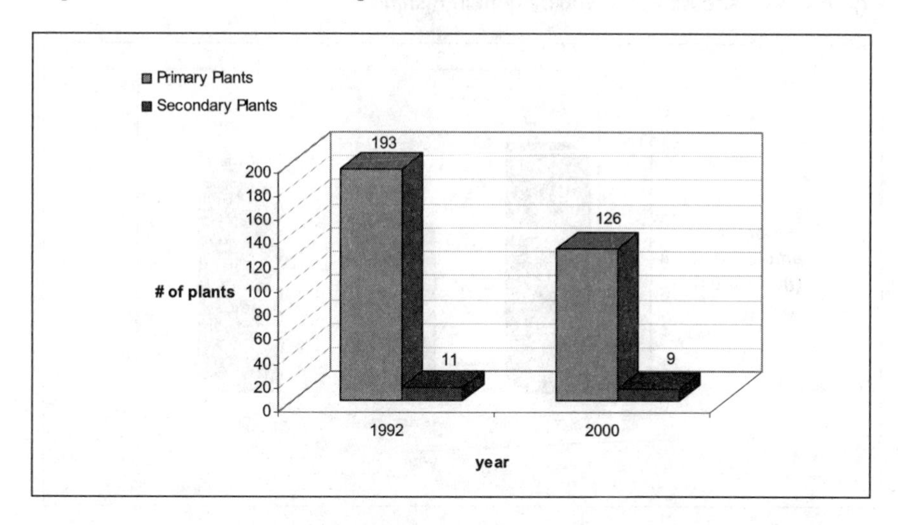


1992 the province licensed 193 primary processing plants and 11 secondary plants; in 2000 there were 67 fewer active licensed primary plants and two fewer secondary processing plants (see Figure 2) (Department of Fisheries and Aquaculture, 2003). "Average monthly employment in fish processing in Newfoundland and Labrador between 1987 and 1998 peaked at 13.3 thousand in 1989, declined to a low of 3.7 thousand in 1994 and increased again to 6.2

Figure 3: Average Monthly Employment in Fish Processing

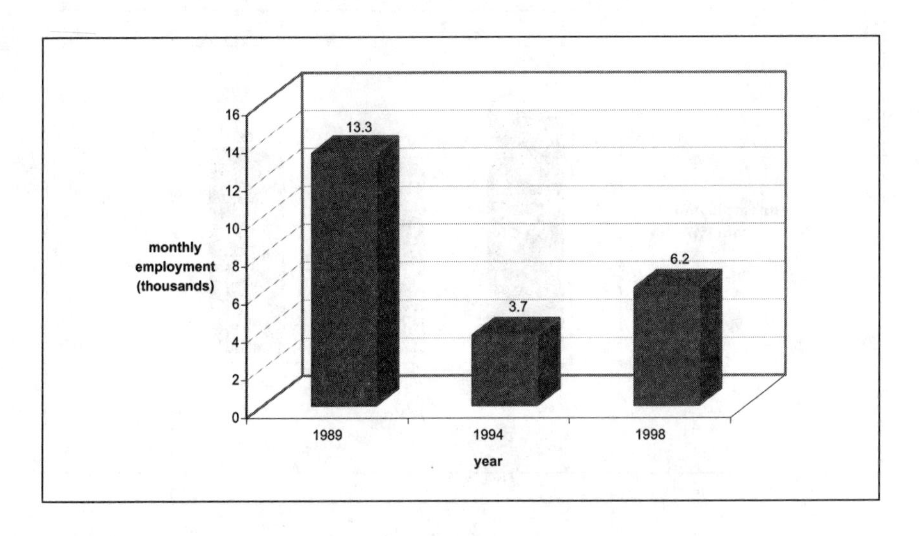


Figure 4: Average Monthly Employment in Fishing

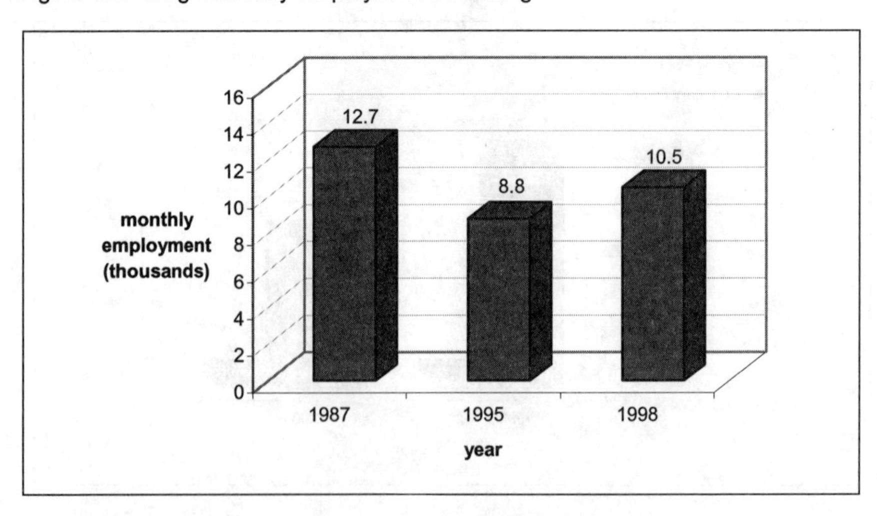


thousand in 1998" (see Figure 3) (Neis et al., 2001: 40). For fishing, "average monthly employment declined from a high of 12.7 thousand in 1987 to a low of 8.8 thousand in 1995, increasing again to 10.5 in 1998" (see Figure 4) (Neis et al., 2001: 40). While the number of people employed in the fishing industry has declined, incomes have increased (see Figure 5).

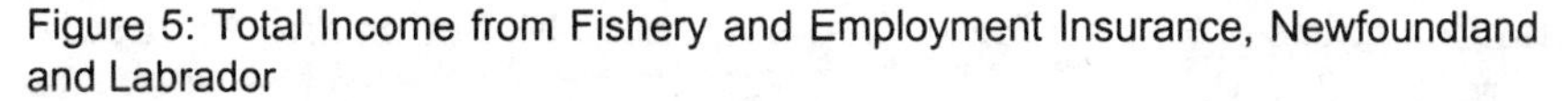

Figure 5: Total Income from Fishery and Employment Insurance, Newfoundland and Labrador

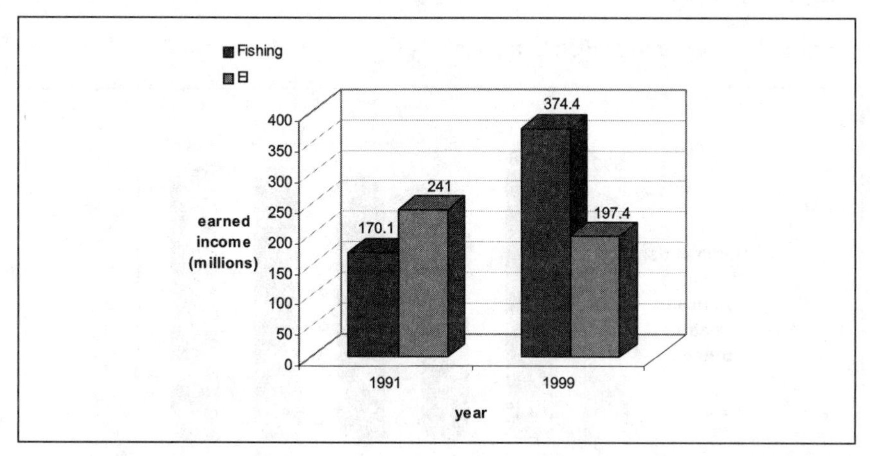


> Taxfiler data for 1991 . . . and 1999 . . . indicates that the number of people deriving income from the fishery declined by 8.4%, whereas total incomes (from all sources) increased by 37.4%. . . . Earned income from fish related activities more than doubled from $170.1 million in 1991 to $374.4 million in 1999. E.I. payments, on the other hand, declined from $241 million in 1991 to $197.4 million in 1999. Fishers' and processors' average annual incomes from all sources increased from $15,046 in 1991 to $22,572 in 1999. . . . [T]here were fewer harvesters and fewer processing workers in lower income ranges in 1999 compared to 1991. Conversely, the number of workers in higher income groups increased considerably. For example, the combined number of harvesters and processors with incomes of $30,000 or more almost triples to over 6,500 in 1999. About 75% of the increase in the high income categories accrued to fishers. On an overall basis, over 81% of fishery workers in 1991 had incomes of $20,000 or less; by 1999, those with incomes of $20,000 or less accounted for only 55% of the total. (Economics and Statistics Branch, 2003: 19)

Increased incomes reflect increased value of fish landings and the increased concentration of wealth. Between 1992 and 2002, the total volume of fish landings in Newfoundland and Labrador reached a low of 127,500 metric tonnes in 1994, peaked at 271,500 metric tonnes in 1999, and fell to 261,000 metric tonnes in 2001 – still lower than 278,000 metric tonnes in 1992 (see Figure 6). Yet the total value of fish landings increased in the same period, from $179.3 million in 1992 to $570.7 million in 2000, dipping in 2001 to $487.2 million (see Figure 7) (Government of Newfoundland and Labrador, 2003). Total landings in 2002 were 267,470 tonnes with a landed value of $515 million (Department of Fisheries and Aquaculture, 2003b: 1).

Figure 6: Total Volume of Fish Landings

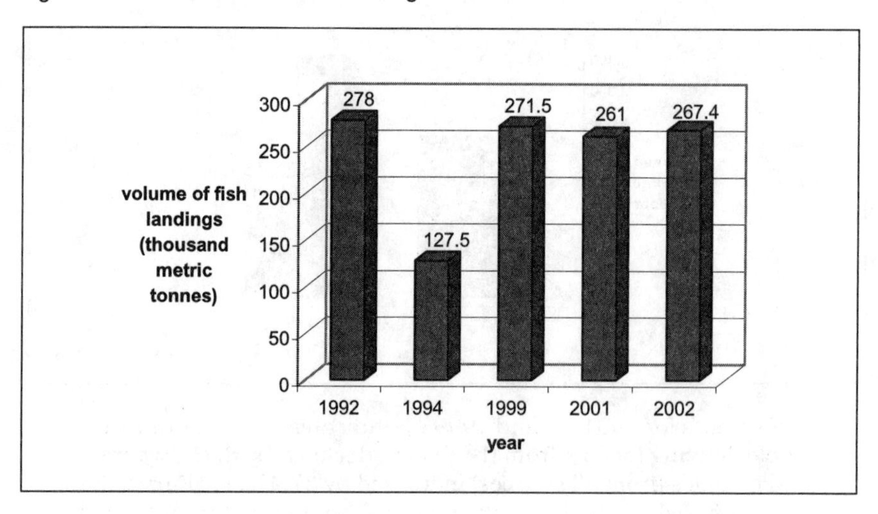


Figure 7: Total Value of Landings

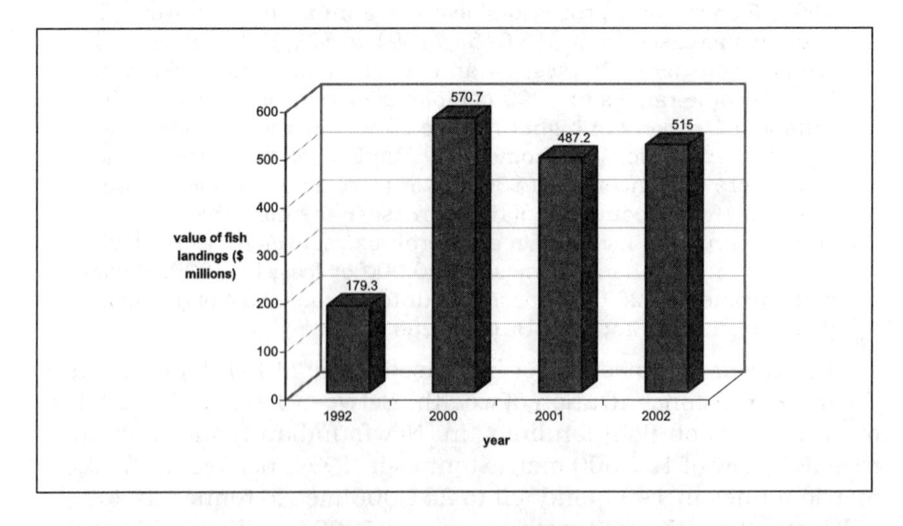


In 2002, the production value of the fishing industry in Newfoundland and Labrador exceeded $1 billion – the third time since 1999 – reflecting an industrial concentration on shellfish and high prices for crab (Department of Fisheries and Aquaculture, 2003b:

1). Regional and scale dynamics have mediated the impacts of moratoriums and industrial restructuring and have reshaped access to fisheries wealth. Regions with a recent history of crab and shrimp harvesting and processing are coping better than those heavily dependent on the groundfishery. In 1990 groundfish landings were 336,588 metric tonnes, with a landed value of $174 million; in 1993, 97,650 metric tonnes were landed, with a value of $50.9 million; and in 2001, 67,421 tonnes were landed, with a value of $68.5 million (see Figure 8) (Department of Fisheries and Oceans, 2003d). Shellfish landings, on the other hand, have increased from 47,495 tonnes in 1990 with a landed value of $76.4 million to 152,368 tonnes in 2001, valued at $431.2 million (see Figure 9) (Department of Fisheries and Oceans, 2003d). In 1988, shellfish species accounted for less than 30 per cent of the total landed value; in 2002, they made up 82 per cent of total landed value (Department of Fisheries and Aquaculture, 1998; Department of Fisheries and Aquaculture, 2003b: 2). Crab has been a particularly lucrative species in recent years. In 2001 crab landings were 58,107 tonnes (with a landed value of $219.9 million), up from 11,054 tonnes in 1990 (with a landed value of $13 million) (see Figure 10) (Department of Fisheries and Oceans, 2003d).

Figure 8: Groundfish Landings and Landed Value

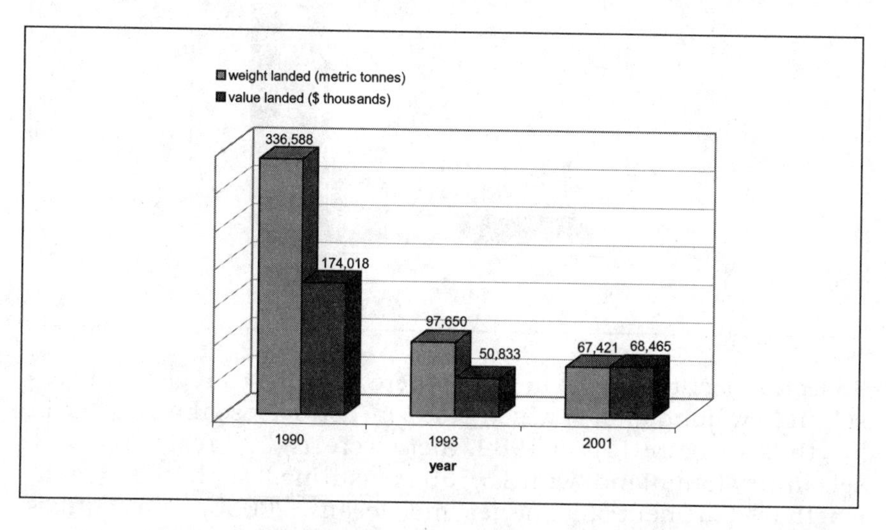


This shift in species has been particularly detrimental for the small boat sector, especially those without access to crab permits, and thus, without access to fisheries wealth. Steady declines from 1989 to 1999 in the total number of vessels, especially those under

Figure 9: Shellfish Landings and Landed Value

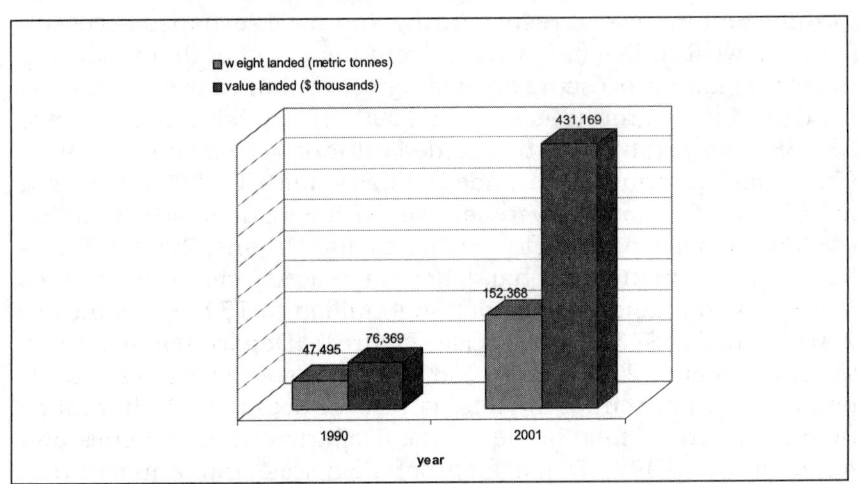


Figure 10: Crab Landings and Landed Value

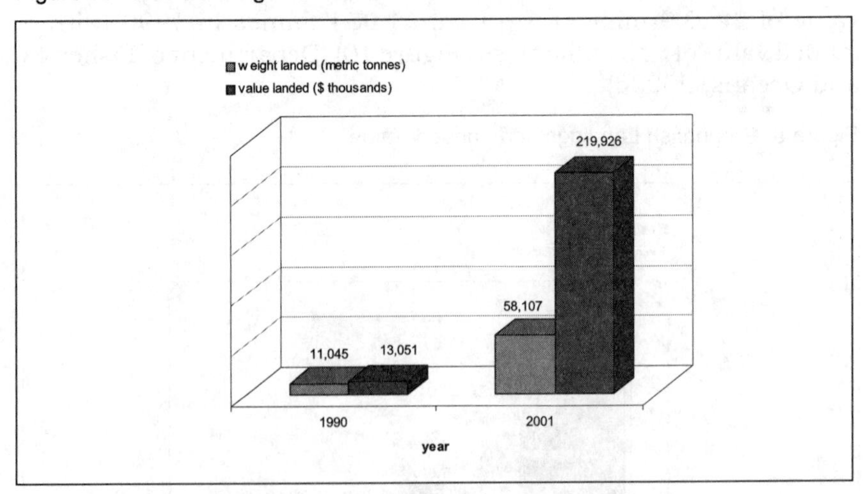


35 feet, reflect this shift. In 1999, there were 9,613 registered vessels in Newfoundland, of which 8,605 measured less than 35 feet in length (see Figure 11). In 1989, there were 17,027 registered vessels in Newfoundland, with 15,730 measuring less than 35 feet in length (Department of Fisheries and Oceans, 2003b). The acquisition of crab licences is increasingly necessary for fisheries success, and except for the 1993-94 period there has been a steady increase in the number of crab licences allocated in Newfoundland and the total Atlantic region. In 2000, 779 crab licences were issued in

Figure 11: Licensed Vessels

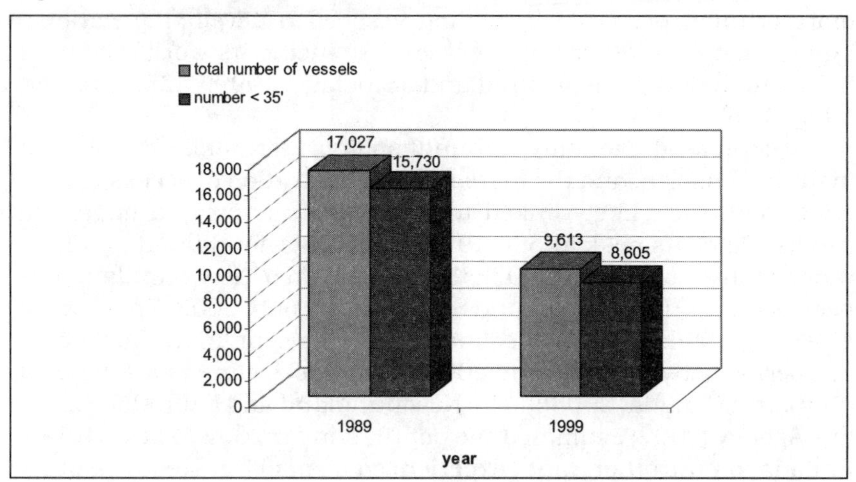


Figure 12: Crab Licences Issued, Newfoundland and Atlantic Region

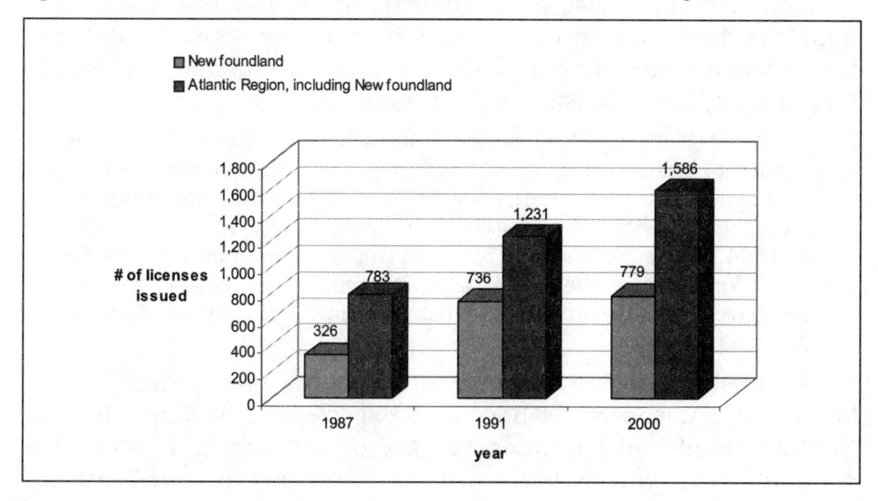


Newfoundland and 1,586 in the total Atlantic region; in 1991, 736 licences were issued in Newfoundland, up from 326 in 1987, and 1,231 licences were issued for all of the Atlantic region, up from a total of 783 (see Figure 12) (Department of Fisheries and Oceans, 2003a). In 1999 there were 778 crab licences issued in Newfoundland and Labrador, and approximately 3,400 inshore enterprises participated in the crab fishery (Department of Fisheries and Aquaculture, 2003; Department of Fisheries and Oceans, 2003). The dominant processing company in Newfoundland, FPI, has experienced a less volatile situation. FPI has maintained profits despite the moratoriums, primarily by selling vessels to developing nations, buying fish from other suppliers, and reducing its workforce from 8,000 to 2,000 in Newfoundland (Sinclair, 1999: 325; Ommer, 1995: 175).

A depressed economy, groundfish moratoriums, employment insurance reform, and the increased concentration of access to fisheries wealth are likely linked to recent population and migration trends. Since its peak at 580,195 in 1993 (Newfoundland and Labrador Statistics Agency, 2002), the population of Newfoundland has been declining (Economics and Statistics Branch, 2003: 7). Between 1991 and 1996 the rate of decline was 2.9 per cent; in the latest census period between 1996 and 2001 the rate of decline was 7 per cent (Statistics Canada, 2003b). The Newfoundland and Labrador Statistics Agency (2002) estimated the population for 2002 to be 531,595. Canada, on the other hand, experienced a small but steady population increase of 4 per cent between 1996 and 2001 (Statistics Canada, 2003b). These population trends reflect that Newfoundland has the strongest out-migration, which peaked in 1998, and lowest fertility rates among all the provinces (Economics and Statistics Branch, 2003: 7; Statistics Canada, 2003b).

> Over the 1994 to 2002 period, falling fertility rates and fewer women of childbearing age caused births to decline while population aging resulted in more deaths. . . . The natural [population] increase was only 269 in 2002, down dramatically from over 2,400 in 1994. These trends have resulted in rapid population aging. In 2002, for example, the median age of the population was 38.5 years compared to 32.5 years in 1994. (Economics and Statistics Branch, 2003: 7)

While those areas heavily dependent on the groundfishery were hardest hit, urban areas have also experienced declines. In fact, Newfoundland and Labrador was the only province in which the population of urban centres had declined (Economics and Statistics Branch, 2003: 7; Statistics Canada, 2003b). These trends will have a negative impact on the local economy as younger and better educated people leave and equalization payments are reduced (Human Resources Development Canada, 1998: 77, 79).

## THE STUDY AREA

The Bonavista-Trinity Bay region of Newfoundland is the home of the people in this study (see map). According to 2001 census information, over 12,700 people lived in this area – down from approximately 14,700 in 1996. At least 19 communities are represented in this research, and each has experienced population declines between 1991 and 2001 (Statistics Canada, 2001, 1996). The standard of living in this area appears similar to that of other rural areas in Canada (Ommer, 1998: 48). The General Social Survey and Community Ecology Section of the Eco-Research Project (ibid., 47-9) report that both the Isthmus and the Bonavista Peninsula areas of the study region are similar in a number of ways. Both areas have a high rate of home ownership (90 per cent), a similar mean household size (3.57 and 3.54 respectively), and a disproportionate reliance on seasonal jobs (upwards of 50 per cent), and are marked by a rigid gender division of labour and heavy dependence on transfer payments. Both areas also share a similar occupational distribution of the labour force – harvesters and processing workers and other blue-collar jobs making up the majority (approximately 60 per cent) of jobs held, followed by employment in service, clerical, and sales (between 25 and 30 per cent) and professional and technical jobs (between 10 and 15 per cent) (ibid., Table G9). Residents in both areas reported a general satisfaction with their standard of living and the quality of life experienced, and they expressed an attachment to the area in which they lived.

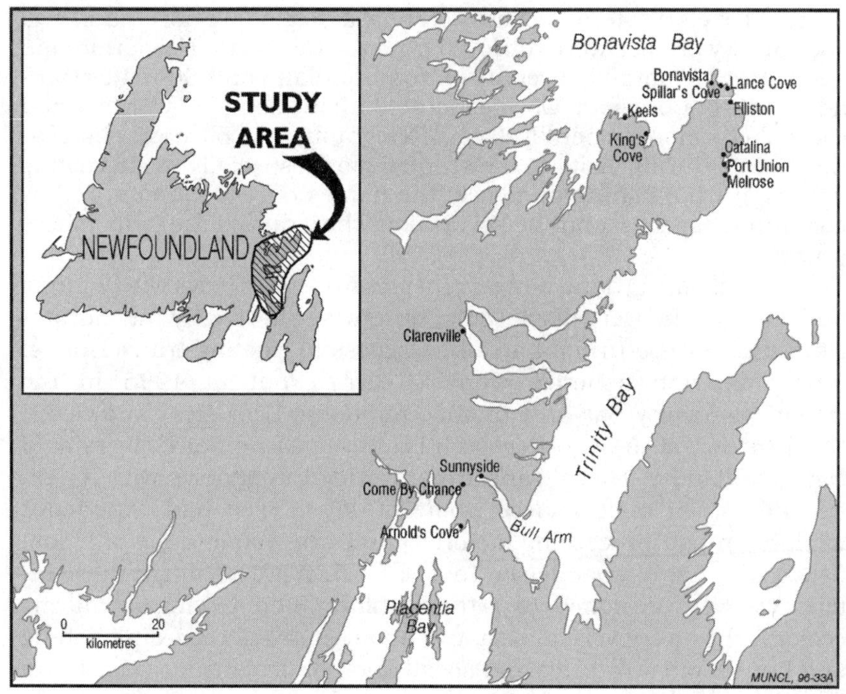

Map of Study Area

Differences between the Isthmus and the Bonavista Peninsula areas reflect specific historical and geographical patterns of resettlement and industrial development (ibid., 47-9). Bonavista Peninsula residents were more likely than Isthmus residents to have roots in local fishing families. Isthmus residents were more likely to be employed and enjoy higher total incomes ($29,898 for Isthmus residents and $16,637 for Bonavista residents) and employment incomes ($10,975 for Isthmus residents and $5,780 for Bonavista residents). Yet Isthmus residents were more likely to perform informal work for or without pay. People on the Isthmus were also more likely to have post-secondary education and less likely to be poorly educated.

## DATA COLLECTION

The data used in the present research include documents and official Web sites produced by the government of Canada, including the Department of Fisheries and Oceans and Statistics Canada, by the government of Newfoundland and Labrador, including Newfoundland and Labrador Statistics Agency, the Department of Fisheries and Aquaculture, and the Economics and Statistics Branch, by the Fish, Food and Allied Workers, including the *Union Advocate*, and by the Canadian Council of Professional Fish Harvesters and the Professional Fish Harvesters Certification Board. Regional information and community profiles were retrieved from the Centre for Newfoundland Studies at Memorial University of Newfoundland and Statistics Canada and from previous work generated about fisheries workers and the fishery more generally (see Bibliography). I collected observational data during visits to the study region, especially while visiting fishers' private fishing premises, the homes of respondents, public community spaces, and the Port Union wharf during the crab fishery season.

The bulk of my data, however, comes from interviews with 71 men and women who were displaced or otherwise affected by the moratorium and restructuring. I acquired access to existing transcripts of interviews with fisheries workers carried out in 1995 in the Bonavista-Trinity Bay area by the Traditional Ecological Knowledge (TEK) section of the Eco-Research Project at Memorial University of Newfoundland.[11] These transcripts provided interviews with 41 respondents: 39 male fishers (some of whom also had experience working in fish-processing factories) and two women (one of whom fished with her husband) (see Table 1 for details). Although these interviews were designed to retrieve information regarding marine ecology, they were very informative about changes regarding fishing and household strategies and social and political issues relevant to the lives of fishery peoples.

Table 1: Interviews Conducted by the TEK Project, Memorial University of Newfoundland

| ID# | Date | Occupation | Sex | Age[i] |
|---|---|---|---|---|
| TEK-3[ii] | 20/6/95 | Fisher | Male | 46 |
| TEK-5 | 21/6/95 | Fisher | Male | 60 |
| TEK-7 | 22/6/95 | Fisher | Male | 76 |
| TEK-8 | 11/7/95 | Fisher | Male | 69 |
| TEK-10 | 11/7/95 | Fisher | Male | 31 |
| TEK-11 | 27/6/95 | Fisher | Male | 45 |
| TEK-12 | 4/7/95 | Fisher | Male | 65 |
| TEK-13 | 5/7/95 | Fisher | Male | 44 |
| TEK-14 | 6/7/95 | Fisher | Male | no data |
| TEK-15 | 6/7/95 | Fisher | Male | 37 |
| TEK-16 | 5/7/95 | Fisher | Male | 70 |
| TEK-17 | 10/7/95 | Fisher | Male | 73 |
| TEK-18 | 12/7/95 | Fisher | Male | 39 |
| TEK-19a[iii] | 12/7/95 | Fisher | Male | 42 |
| TEK-19b | 12/7/95 | Fisher | Female | no data |
| TEK-20 | 12/7/95 | Fisher | Male | 30 |
| TEK-21 | 13/7/95 | Fisher | Male | 65 |
| TEK-22 | 14/7/95 | Fisher | Male | no data |
| TEK-23 | 19/7/95 | Fisher | Male | no data |
| TEK-24 | 19/7/95 | Fisher | Male | no data |
| TEK-26 | 20/7/95 | Fisher | Male | 42 |
| TEK-27 | 21/7/95 | Fisher | Male | 46 |
| TEK-28 | 21/7/95 | Fisher | Male | 40 |
| TEK-29 | 24/7/95 | Fisher | Male | 85 |
| TEK-31 | 25/7/95 | Fisher | Male | 40 |
| TEK-32 | 25/7/95 | Fisher | Male | 50 |
| TEK-33 | 25/7/95 | Fisher | Male | 33 |
| TEK-34 | 26/7/95 | Fisher | Male | 73 |
| TEK-41 | 6/11/95 | Fisher | Male | 52 |
| TEK-43 | 7/11/95 | Fisher | Male | 52 |
| TEK-44 | 8/11/95 | Fisher | Male | 32 |
| TEK-45 | 8/11/95 | Fisher | Male | 48 |
| TEK-46 | 9/11/95 | Fisher | Male | 38 |
| TEK-47 | 9/11/95 | Fisher | Male | 44 |
| TEK-48 | 10/11/95 | Fisher | Male | 42 |
| TEK-50 | 12/11/95 | Fisher | Male | 50 |
| TEK-51a | 12/11/95 | Fisher | Male | 77 |
| TEK-51b | 12/11/95 | Salt-Fish Maker | Female | no data |
| TEK-53 | 13/11/95 | Fisher | Male | 45 |
| TEK-54 | 14/11/95 | Fisher | Male | 44 |
| TEK-55 | 14/11/95 | Fisher | Male | 58 |

[i] This heading refers to the age of the respondent at the time of the interview.
[ii] The numbers assigned here to each respondent correspond to the number assigned to each interview transcript by the TEK project.
[iii] Marital relationships are indicated by a common number, followed by an *a* or *b*.

I interviewed the remaining 30 respondents in 1997 and 1998 (see Table 2 for details). Interviews took place either in the homes of respondents or in community spaces such as local restaurants. While the focus of this book is male small-boat fishers, I interviewed both men and women because of the relational nature of the local constructions of gender, as well as to capture not just men's but women's more marginal experience of patriarchy and masculinity. I

Table 2: Interviews Conducted by Author

| ID# | Date | Occupation | Sex | Age |
|---|---|---|---|---|
| NGP-1 | 25/11/97 | Fisher | Male | 35 |
| NGP-2 | 4/11/97 | Fisher | Male | 40 |
| NGP-3 | 3/11/97 | Fisher | Male | 71 |
| NGP-4a | 23/7/98 | Fisher | Male | 39 |
| NGP-4b | 23/7/98 | Fisher | Female | 37 |
| NGP-5 | 14/7/98 | Fisher | Male | 33 |
| NGP-5b | 14/7/98 | Fisher | Female | no data |
| NGP-6a | 7/6/98 | Fisher | Male | 86 |
| NGP-6b | 7/6/98 | Salt-Fish Maker | Female | no data |
| NGP-7a | 7/6/98 | Fisher | Male | 46 |
| NGP-7b | 7/6/98 | Non-Fisheries | Female | no data |
| NGP-8 | 14/6/98 | Fisher | Male | 40 |
| NGP-9a | 15/6/98 | Fisher | Male | 44 |
| NGP-9b | 15/6/98 | Fisher | Female | 44 |
| NGP-10 | 4/11/97 | Plant Worker | Male | 39 |
| NGP-11 | 24/11/97 | Plant Worker | Male | 36 |
| NGP-12a | 24/11/97 | Plant Worker | Male | 57 |
| NGP-12b | 24/11/97 | Plant Worker | Female | no data |
| NGP-13a | 5/11/97 | Plant Worker | Male | 58 |
| NGP-13b | 5/11/97 | Plant Worker | Female | 49 |
| NGP-14a | 15/6/98 | Plant Worker | Male | 41 |
| NGP-14b | 15/6/98 | Plant Worker | Female | 40 |
| NGP-15a | 23/9/98 | Plant Worker | Male | 43 |
| NGP-15b | 23/9/98 | Plant Worker | Female | 39 |
| NGP-16a | 22/9/98 | Plant Worker | Male | no data |
| NGP-16b | 22/9/98 | Plant Worker | Female | 51 |
| NGP-17 | 3/11/97 | Plant Worker | Female | 48 |
| NGP-18 | 14/6/98 | Plant Worker | Female | 53 |
| NGP-19 | 22/9/98 | Plant Worker | Female | 51 |
| NGP-20 | 7/6/98 | no data | Female | no data |

also interviewed men and women who worked in local fish-processing plants in order to separate gender effects from effects based on positions within production relations. At times, I interviewed husband-wife couples simultaneously; other times, I interviewed one of the partners in the absence of the other.

I employed a qualitative, open-ended interview approach to my research for a number of reasons. This approach allows respondents to exercise some control over the interview process and is flexible, enabling changes to be made to the interview schedule (Collinson, 1992: Appendix; Judd et al., 1991: 239-40, 261). I often added relevant issues as they were brought to my attention and discarded questions that were simply irrelevant to the lives of respondents. I asked a series of questions about paid and unpaid work histories, especially in relation to the fishery and to personal and household situations since the cod moratorium was declared. Once the interviews had been completed and transcribed, I started my analysis by looking for significant patterns, especially with regard to age, sex, and job, as well as evidence of individual and collective behaviour and ideas. Leach (1997: 43) says that interviews demonstrate how behaviour and ideas are often contradictory, which "is important in political terms for pointing to the spaces potentially open to counterhegemonic strategies." Neysmith (2000: 13) writes that qualitative methods "offer more conceptual space than do quantitative research approaches for alternative voices to be heard." Qualitative methods have enabled me to use excerpts from interview transcripts to remind the reader that *real* people are experiencing what is described in this book. Of course, reproducing only words cannot always convey the sentiments communicated through facial expressions and gestures. Some excerpts have been edited for style and to protect the identity of the respondents.

## THE PEOPLE IN THE STUDY

Generally speaking, most of the 71 respondents come from families and communities that have traditionally depended on the fishery, directly or indirectly. Fifty-five (or 77.5 per cent) of the respondents are men – 48 fishers and seven plant workers. Sixteen (22.5 per cent) of the respondents are female – four fishers, eight plant workers, two salt-fish makers, and two not directly involved in fisheries work. Respondents tended to be married or cohabiting with common-law partners and only two (3 per cent) respondents reported having no children at the time of the interviews.[12] Out of the 48 male fishers, at least nine had spouses who fished with them at one time or another, and at least 20 of the spouses were identified as processing workers – nine of whom were salt-fish makers. All of the female fishers crewed in boats with male partners. All of the spouses of the seven male plant workers were plant workers. Out of the eight female plant workers, six had spouses who worked in the plant and two were married to fishers. Of the 15 plant workers I spoke with, only two, both

The FPI Bonavista fish plant (Photos by Nicole Gerarda Power).

men, reported getting occasional but yearly work at local fish plants since 1992 and one had retired. Two of the four women fishers have withdrawn from the harvesting sector – one due to injury, the other as a result of obstacles following restructuring.[13] However, this latter withdrawal may be temporary.

Eleven of the fisher respondents reported being retired. Five of these retired male fishers reported the moratorium as the reason for retirement. Only two of the 37 active male fishers reported not fishing at all, although most reported spending substantially less time fishing since the moratorium. Ages of respondents ranged from 30 to 86 at the time of the interviews. My failure to include men under the age of 30 reflects the trend of out-migration among younger men, which is likely to have been exacerbated by the fisheries crisis. Throughout this book I use the terms "older" and "younger" to differentiate between fisheries workers aged 55 and over and those under the age of 55. I have chosen this arbitrary way of talking about age differences among fisheries workers because it made sense to respondents in light of the early retirement packages available to those aged 55 and over. This construction demonstrates that age is not necessarily or always a quantitative variable that can be employed objectively to order data.

Forty-seven (or 90 per cent) out of 52 fisher respondents, male and female, participated primarily in the inshore fleet sector, which employs small boats or longliners measuring under 35 feet in length. Fishers often use the terms "inshore" and "offshore" or "small boat" and "big boat" loosely to position themselves within a larger discussion regarding the development of the fishery and to draw attention to different philosophies and approaches to fishing embedded in these sectors. However, it seems that this process is increasingly complicated by the government's neo-liberal direction and is undergoing some significant changes. The work experience of four of the fisher respondents was limited to vessels measuring over 35 feet, two of whom fished as crew on FPI vessels. However, in most instances the male fishers had varied fishing histories, moving between sectors and boat sizes – inshore (vessels under 35 feet), nearshore (vessels between 35 and 65 feet), midshore (between 65 and 100 feet), and offshore (vessels over 100 feet). Such moves mean that most male fisher respondents have acted as crew or owner at different points in their careers and have worked on small open boats requiring much manual labour as well as larger vessels equipped with modern technology. Some of these men also reported previous experience working in fish-processing factories. Forty-one out of 48 male fishers reported being skippers and owning their fish-

The FPI Port Union fish plant (Photos by Nicole Gerarda Power).

ing boats and gear or jointly owning the property with a partner. All four of the women fishers were considered crew in their husbands' enterprises. Some older fishers reported involvement in the household-based salt fishery immediately before and after Confederation. For most fishers, cod had been their bread and butter. However, at least two of the male respondents were exclusively full-time crab fishers.

As stated earlier, I included men and women fish-processing workers in my interview process. While there were a number of processing plants in the research area, the plant workers in this study worked either in the factory located in the town of Bonavista, which was supplied by inshore fishers, or the Port Union factory supplied by offshore vessels. Both plants were owned and operated by the multinational company, Fishery Products International, and the workers were unionized. The inshore plant operated on a seasonal basis and, prior to the moratorium on northern cod in 1992, processed primarily codfish and snow crab. The focus shifted almost exclusively to snow crab after 1992. Some of the inshore plant workers had worked processing only groundfish; others had processed groundfish and crab during their working careers. The offshore plant operated virtually year-round, processing a variety of groundfish species, with a particular focus on cod, prior to the closure of all of these fisheries and the plant itself in 1992. In 1998, the offshore plant reopened as a highly automated, specialized shrimp-processing facility, employing approximately one-tenth of its former workforce.

# *Coping with Ecological Crisis and Adjustment* 3

“THERE’S NOTHING THERE FOR YOU TO CATCH”

The price of “progress” has been high for Newfoundlanders. The declaration of moratoriums on northern cod in 1992 and other groundfish stocks in subsequent years represents a crisis in production based on ecological devastation. Moratoriums and declining stocks have placed tremendous pressure on fisheries workers in terms of making a decent living and have directly altered their work experiences. Indirectly, local businesses and local people in non-fisheries work are also feeling the economic pinch of the current fisheries crisis. Processing workers have been especially hard hit (Neis et al., 2001: 40). Many respondents agreed with this assessment: “[E]verybody is on this TAGS, which is a good thing. If there was no TAGS, people would be in . . . especially the plant workers, because fishermen are still making a few dollars at the lump [a dark, round fish harvested primarily for roe], some at the crab, and lobsters, but the plant workers are out in the cold” (TEK-23, male fisher). Women plant workers seem to have fared worse than men because of the gendered organization of work – within the industry and within processing – and of seniority lists.

Many fishers, on the other hand, have continued to work by shifting effort from cod to other species for which they already have licences. Other fishers have diversified and acquired new licences. Nonetheless, the amount of time spent on water has been drastically reduced for most of the fisher respondents, meaning they spend more time on land and at home than in previous years. Moreover, making a decent living from fishing alone has become impossible for many men, particularly those fishing from small boats in the inshore: "You can't make a good living now . . . because there's nothing there for you to catch" (NGP-9a, male fisher). At the time of the interviews, some fishers reported that fishing time had been reduced from four or six months to six weeks. Some inshore fishers decided to withdraw from particular fisheries for financial reasons. Participation in such open fisheries as lump often entails an economic loss due to an imbalance between the costs of harvesting the species and the prices offered to fishers for the raw material. Fishers also described problems they encountered in acquiring quotas or substantial quotas for such particularly lucrative species as crab. This is not to suggest that all fishers have lost since the moratorium – in fact, some have gained, earning more now than ever before. This reflects an apparent wider trend – one with gender and class consequences – of increased concentration of access to resources and wealth in the fishery.

Aside from the financial compensation component of the NCARP and TAGS programs, most of the state's initiatives that have accompanied the moratoriums on northern cod and other groundfish species – downsizing and adjustment into non-fisheries-related jobs – have encountered resistance. Behind these initiatives lie certain assumptions. One assumption is that there are too many fishers and not enough fish. This view ignores the differential impact of the various technological and harvesting capacities between the inshore and offshore sectors. A second assumption is that displaced fisheries workers will be willing to participate in retraining and that there will be jobs available after retraining. Finally, embedded in state initiatives is the assumption that people will be willing to fill these jobs, even if this entails moving from their communities. These assumptions are unfounded.

Male and female respondents tended to interpret state initiatives accompanying the moratoriums, including financial compensation, early retirement packages, retraining opportunities, and a process of buying back fishing licences, as attempts to remove people from the so-called "backward" traditional fishery. The state's focus on downsizing has exacerbated fears of exclusion in the future fishery among inshore fishers: "Small boat haven't got a chance. . . . Anything less than thirty-five footer, I think they wants to give the boot anyway" (NGP-4a, male fisher). Similarly, fishers pointed to feelings of marginalization and an awareness of a loss of power: "The problem now is in numbers of fishermen and plant workers. It is going to come to the point where there is not going to be many left after the moratorium. Then you're not going to have much say at all" (TEK-11, male fisher). Local fisheries workers, male and female, generally supported these assessments. Nonetheless, most respondents were grateful for having received financial compensation through NCARP, and later TAGS, which enabled them to stay in their communities. At the same time, many perceived state financial compensation as a form of social control, intended to pacify fisheries workers and limit resistance to state policies and fisheries closures: "We always thought that this bit of money we's getting on TAGS was keeping the people shut up, you know?" (NGP-1, male fisher).

Resistance to outside directives pertaining to the fishery is nothing new.[1] Not surprisingly, then, *one* of the ways – and the most common – in which fishers responded was with resistance towards state-directed initiatives and removal from the fishery. There are a number of obvious reasons why fisheries workers would resist removing themselves from the fishery and their communities, as well as some less obvious ones. I argue that the various coping strategies and responses developed by male fishers to deal with their circumstances since the moratorium are mediated by gender. Sexual divisions of labour in the fishing industry and fishing households, and the gender ideologies that maintain them, organize how women and men cope with restructuring, and make some problems and possible solutions more relevant or available to men than women and vice versa. In this case, men fishers coped by using available discourses and material resources, which reflect and reproduce particular gender configurations that are culturally specific. These gender configurations reflect male dominance in the fishery, their contradictory class position, and the local cultural importance ascribed to fisheries work as a space for defining manhood. In this chapter I describe and analyze perhaps the strongest and most common response to restructuring – "looking to the past" and drawing on the "traditional male fisher" model. However, exactly how this type of resistance was manifested and the precise meanings attached to such resistance varied among fishers.

## "WE'RE STILL FISHING"

While some fishers have been edged out of the fishery, however slowly, through early retirement schemes and adjustment initiatives, others have been able to resist abandoning investments because limited work with other lucrative species has offset some of the financial burden. In other words, fishers have been able to resist removal from the fishery largely by choosing to harvest species other than cod and other banned groundfish – despite clawback rules attached to compensation packages that deduct fishing earnings from

TAGS income (Fish, Food and Allied Workers, 1996: 13, 19). Why? Because they can. And many feared that if they failed to demonstrate such commitment to fishing they would be disadvantaged by downsizing policies designed by the state to eliminate as many fishers as possible. State directives such as these and the "use it or lose it" approach of DFO to licences mean that fishers also experienced pressure to use licences for declining species, despite ecological concerns (Fish, Food and Allied Workers, 1994a: 8). Others have decided that fishing is simply no longer a feasible strategy, especially when NCARP and TAGS provided greater incomes than fishing.

Crab boats docked in Catalina (Photo by Nicole Gerarda Power).

Stock declines and moratoriums have necessitated that fishers wishing to remain in the fishery adopt a number of fishing strategies. Men made decisions about which strategies to pursue, not as isolated individuals but as socially embedded actors, considering household and crew needs as well as available resources. Some fishers moved into smaller boats requiring less crew and gear or began to fish with their wives, often depending on the age of children and the availability of child care. Less investment and husband-wife crews allowed more potential income to go into the household. However, the establishment of government policies that eliminate part-timers and the creation of a gradient fishing status system may

Ready for lobster season (Photos by Nicole Gerarda Power).

encourage women to remove themselves from the harvesting sector (Neis, 1993: 204). Some fishers have decided to fish on their own since the moratorium: "Since the moratorium came into effect . . . there is not enough money in it for two people. So we had to part ways, and fish by ourselves. It's not a good safety policy, but what can you do? . . . It makes everything more difficult and harder to work, but you've got to put up with that" (TEK-31, male fisher).

Another strategy developed by fishers to resist removal from the fishery included forming co-operative work relationships with other fishers, reflecting a willingness on the part of fishers to let go some of their "secret" fishing and ecological knowledge. It appears that the economic imperative to share knowledge with fisher peers outweighs the cost of losing family fishing and ecological "secrets": "Before, everybody used to be for themselves. . . . So there was six of us, and we all buddied up here . . . when I started fishing first, the people over there, they were the bad people . . . over there coming over and taking over our shore type of thing. . . . So since that [moratorium], it changed" (TEK-44, male fisher).

Another, although less popular, strategy to deal with resource scarcity and changing economics in the years preceding and immediately after the 1992 moratorium involved increasing financial investment into the fishery. This approach has meant purchasing communication and labour-saving technology and gears and larger boats to catch more fish. Such purchases provided opportunities to spend more time on the water, increase harvesting effort, and diversify. According to small-boat fishers, new state policies that favour larger boats in terms of allocating larger quotas and licences for other species have placed tremendous pressure to choose between "going bigger" or getting out. However, moving into larger boats is difficult due to restrictive DFO requirements and a lack of financial resources. The two main strategies – going bigger or scaling down – also reflect and reinforce contradictory class-related orientations to the fishery.

### "WHEREVER YOU CAN GET A BUCK"

Not all tactics designed to resist removal and maximize household income were as mindful of the law. Men also manipulated the welfare system, participated in the informal economy, and fished illegally for the purpose of selling or personal consumption – although the extent to which they did so is difficult to gauge. Deceptive and illegal activity is not necessarily new behaviour – it appears, for example, that misreporting catches in the 1980s and 1990s was not particularly uncommon – and therefore cannot be explained simply by current economic circumstances. Nonetheless, unrecorded work for pay appears to be widespread as TAGS recipients could sell their skills cheaply (Sinclair, 1999: 332; 2002: 312). Many fishers argued that being honest with the government or playing by the rules has not served them well. For example, clawback regulations embedded in state financial compensation packages that required fisheries workers to pay back compensation earnings after having surpassed a defined income threshold served as a disincentive to declare earnings (Fish, Food and Allied Workers, 1996: 13, 19). Some respondents suggested they are "forced" to fish and sell illegally to make a living. Fisheries mismanagement and disadvantageous policies have fostered a "screw the system" approach as one fisher put it (NGP-7a, male fisher). Fishers recounted numerous examples demonstrating how playing by the rules has failed them and claimed that desperate times justify these efforts to survive. Government attempts to stop fishing activity become irrelevant when there are no other opportunities to make a living: "As long as you got a boat, you got a job, be it legal or illegal" (NGP-7a, male fisher).

Illegal actions and manipulation of state rules were defended in terms of economic need. Such activities were also clearly symbolic displays, intended to demonstrate a lack of respect for authority and to resist state control over their lives. These activities, though, are unlikely to effect real change. Fishers resisted because fishing was supposed to be a job for life. In addition, many Newfoundlanders believe they have a traditional and cultural right to access fisheries resources at least for personal consumption, whether or not they are bona fide fishers. Current restrictions on both commercial fishing and fishing for food were interpreted by some locals as an attack on their way of life or as institutionalized processes that criminalize ordinary Newfoundlanders.

Poaching may reflect a particular perspective on access to fisheries resources – I call this the "looking to the past" perspective. Consistent with this perspective, all Newfoundlanders – which really means all male Newfoundlanders – have the right to fish for food. The fact that not only male fishers but other local men practise poaching supports this idea. One male fisher (TEK-43) suggested that there may have been fewer incidents of poaching by non-fisher locals in the pre-moratorium period. For him, this demonstrates a local respect for fishers and an oppositional stance against authority: "There's a lot of fish brought ashore now, compared to when you were allowed to go out and catch it. I'm not saying for the traps and gill nets, the bona fide fisherman. But just the ordinary Tom, Dick and Harry going out and jigging his fish, there's more brought ashore now than then." Poaching fish has become a way for local men to be engaged in confrontational practices in the moratorium period.

Willis (1977: 34) makes an argument about fighting and thieving among British working-class men that can be extended to incidents of poaching by Newfoundland men. Like fighting or thieving, poaching "breaks the conventional tyranny of 'the rule'" and transforms the normal and the present into situations of the utmost importance and consequence. In other words, there is a certain excitement and sense of immediacy involved in breaking rules: "People like that just go for the thrill. And not only that, everyone is desperate, desperate for a meal of fish and they pay two or three dollars a pound for it. And for those guys, easy dollar. That's a dozen beer. Two fish is a dozen beer" (NGP-9b, female fisher). While the need for cash or simply a meal of fish may be motivation enough, poaching also relieves daily monotony and provides a sense of freedom, as well as a sense of success in having flaunted authority. It becomes an act of symbolic importance. Suspicion and distrust concerning the intentions of the government, reinforced by rumours about undercover fisheries officers in the area, provided the context out of which the opportunity for and interest in poaching arose for many men.

## "EVERYTHING IS COMING BACK HERE AGAIN NOW"

Interviews suggest that strategies of "looking to the past" are not confined to the level of ideas or symbolism, but have a practical purpose. One strategy adopted by local people to deal with financial and social hardships resulting from moratoriums is returning to or relying more heavily on subsistence practices, including cutting wood for fuel and hunting and gardening for personal consumption. Sinclair's (2002: 313-15) findings corroborate this assessment. One male fisher (NGP-9a) reported that in recent years he and his wife have focused much of their time on hunting and raising animals for personal consumption: "Well here . . . everything is coming back here again now." It seems that they are not the only ones to take this direction. While the nature of and time spent carrying out traditional subsistence activities have changed over the years, these practices had never totally been disregarded. Nonetheless, for many, moratoriums reinforced the cultural value attached to the idea of self-sufficiency and the economic importance of subsistence activities.

Of course, fisheries workers have more time now to partake in these activities. Many respondents linked this increased engagement in subsistence activities to both household economics and a need to occupy time: "We still have our own gardens. We grow all our own potatoes. It seems like people starting to get back into that now. . . . We always have potatoes, but three or four years ago I started to raise all this other stuff when I had more spare time, you know? And it was something, it's a hobby that I enjoy" (NGP-10, male plant worker). Contemporary fishers and plant workers expressed pride in having the knowledge and necessary skills to get by, given these terrible circumstances. These strategies are, however, connected to ideas about gender. Given the precarious nature of the fishery, such culturally valued male activities as hunting and woodcutting provide space for men to experience continuity in their lives and space *to be men*. While it was evident that some women participated in hunting and other traditionally male activities, most engaged in subsistence activities regarded as "feminine," including gardening, sewing and mending, and preparing such homemade food items as jam from berries. Clearly, these activities have a more immediate connection to the household.

Equally, other flexible and highly valued survival strategies, including relying on family and personal ties for support, services, and paid and unpaid work, have taken on a particular relevance since 1992. Drawing on kin and friends for help may reflect economic need, as well as a way to maintain close ties during unstable times (Sinclair, 2002: 311). One male plant worker (NGP-10) was unsure as to whether or not familial support counted as a strategy to deal with recent pressures, since such support had always been important in outport Newfoundland: "I suppose in all communities you always had your family to fall back on, right?" These relationship-based strategies are gendered as well. Thus, men exchange such services with other men as assisting in household repairs; women, on the other hand, help other women with child care and domestic responsibilities. Sinclair (2002: 311) reports persistence in inter-household assistance, but a declining persistence. In my work, local estimations of who is doing what and how much depend on who you talk to – with older people tending to think less is being done than younger people.

## "TIME TO GO TO THE WOODS"

Porter (1993: 87) observes that contemporary men fishers from Aquaforte were "bound up less with the occupation of fishing than with the identity of 'being fishermen'." In other words, to be a "fisherman" a man does more than simply fish. The seasonal nature of the inshore fishery, past and present, has rarely provided adequate income for fishers. However, it has granted fishers time to prepare gear, maintain fishing property, and participate in a number of informal, economic, male-specific activities, including hunting, woodcutting, constructing boats and houses, gardening, and raising animals. In the pre-Confederation period, plural adaptations were essential to the rural economy and household survival. The expanded welfare state and the extension of unemployment benefits to male fishers in the post-Confederation period have lessened the immediate need to participate in such informal economic activities. Nonetheless, these male activities remain important and are ascribed much value. According to Porter (1993: 87), fishers regarded time outside of the fishing season not as idle time, and "[f]ar from detracting from their status as fishermen, they, and everyone else, saw it as an advantage to have time to 'go to the woods,' 'to be free,' 'to be your own man'." When not fishing, "being a fisherman" kept one busy: "You're looking at around the first of May to around the end of October, you know? . . . So I mean after that you were on unemployment all winter and then a lot of the guys, they went up the woods, cutting wood, whatever, firewood and whatnot. At the gear, probably spend two months doing up the gear every year too, right? So that kept you busy" (NGP-1, fisher, age 35).

Porter's "being a fisherman" is part of what I call the "traditional male fisher" model and has implications for the construction of masculine identity and practice. Specific informal economic activities and other adaptations have been integrated into fishers' strategies to survive and to be "men." This reflects a historically specific material reality. Fishing households have had to rely on a variety of strategies for survival due to political neglect, economic exploitation, the seasonality of the inshore fisheries, and low earnings. While informal economic practices, subsistence activities, and occupational pluralism have been somewhat disrupted since the 1950s with the introduction of a cash economy and the welfare state that injected cash transfer payments into households, such modern amenities as electricity, which reduced the need to cut wood for fuel, and state policies that restricted alternative employment for fishers, these activities nevertheless remained important in practice and culturally (Neis, 1993: 195-6, 201; Ommer, 1998: 19). For example, one male fisher (NGP-2) suggested that over the years his family reduced the size of the vegetable garden in response to the need to acquire more cash and having less available time, but they never totally abandoned this work. Despite the many social and economic changes since the 1950s, Newfoundland men have been able to actively create spaces in which men could "be fishermen" of sorts. For example, Felt (1987: 28) has reported that men who did not participate in the fishery built sheds, which served as work spaces and male gathering places, in backyards.

The work histories of the men respondents document occupational pluralism and movement in and out of the fishery in order to make a sufficient living. This was a common experience among older fishers who, during the salt fishery period in the pre- and immediate post-Confederation period, had to struggle to survive without state help or unemployment insurance benefits. Retired fishers often pointed to the economic necessity of moving in and out of the fishery before the extension of unemployment insurance to fishers: "Well, when the fishery was over here . . . and whatever you go at, you could get a job, you'd go. Carpentry work or in the woods, whatever you could get. . . . You had to wait 'til the season was finished, and then if it was bad, you had to make up somewhere, wherever you could get a chance to get it" (TEK-51, retired male fisher, age 77). Indeed, men who could find better-paying work often left the fishery entirely. For example, a male fish plant worker (NGP-10) explained that his father moved out of the fishery to take a local land-based job because of economic insecurity with the fishery: "Well I guess it was a better career move for him at the time. The fishery was as always a bit of an uncertainty and he figured it was a better opportunity for him to go to work here, right? Wages were better. It was more secure for him at the time." However, the experience of having to find alternative, particularly land-based, work was not restricted to the oldest respondents. Low earnings necessitated contemporary fishers to find alternative work: "Sometimes we seen bad summers, or like the older fellows say, a poor summer. I remember one year in particular, I had to go to St. John's to try to get enough weeks to qualify for UIC [unemployment insurance benefits], working on construction work" (TEK-26, male fisher, age 42).

Fishers of all ages appreciated the differing generational experiences of surviving rural Newfoundland and recognized the introduction of the welfare state, particularly the extension of unemployment insurance benefits to fishers, as improving their lives and necessary to "get you through the winter months" (NGP-2, male fisher, age 40). Other fishers expressed gratitude for the shift from mercantile relations to a cash-based welfare society: "But I don't want to go back to the blue note.[2] I don't want no merchant owning me" (TEK-26, male fisher, age 42). At the same time, most of the younger fishers perceived the relations of the contemporary struggle to survive as being ultimately the same as they had been in the past. The main difference to them was that the contemporary struggle is played out in more favourable circumstances. The necessity of flexible survival strategies has not disappeared. Rather, they are now employed within the context of and mediated by the welfare state – however precariously – and within a cash economy. In fact, unemployment insurance benefits support the informal economy and make it possible for fishers to successfully practise a seasonal fishery within a cash economy.

Older fishers, on the other hand, tended to use their personal and romanticized histories to prove how the intervention of the welfare state and the adoption of modern technologies have ruined the younger generation's work ethic and strength of character. In this way, older fishers boosted themselves over younger fishers. Hard work was certainly necessary to survive the harsh conditions confronted in rural Newfoundland during the salt fishery period, which was characterized by little government support and an exploitative mercantile system. Older fishers knew all too well the discrepancy between their labour and the price they received for their fish from paternalistic merchants. Hard times instilled a stoic endurance in both men and women. A retired fisher and his wife (TEK-51a/b) discussed the hard times experienced before Confederation, each with reference to gender-specific spaces:

> *Wife*: Sometimes I wonder what we lived on. We never had to go to the government. And we always had a slice of bread to put on the table. And everything was so scarce.
>
> *Husband*: And you get out of the boat when the fishery was over, and take off for the woods, or go carpentry work or something, build a few houses. Or go down in the woods, and even go down in the winter on the pull-off, pulling wood out after it was cut, in the winter pulling it out to the lakes.

This claim to a "hard times past" by older fishers and the perceived relevance of this past articulated by contemporary fishers have reinforced the cultural value ascribed to craft autonomy, hard work, and self-sufficiency. Self-sufficiency takes form as the ability and know-how to live off the land and ocean. One retired male fisher expressed pride in his ability to avoid purchasing meat and vegetables: "We used to sow our potatoes and turnip and carrot and that, still sow them today. We don't have to buy anything like that you know?" (NGP-3). Men also expressed self-sufficiency in their claims of never having gone on welfare, not relying on handouts,[3] and not owing money – sometimes to bolster themselves over others: "Don't get me wrong. There's people got more gear than [me]. But how much money do they owe?" (TEK-43, male fisher). One retired male fisher (TEK-16) was proud of never having acquired a fisheries loan to support his involvement in the fishery: "No, never in my life. Never got that much, big as my thumbnail. . . . Our inshore boats, our small boats, built them ourselves. Our outboard motors, bought they. And we've had twenty of them." Being self-sufficient and debt-free was highly applauded: "And there's say seven I believe is fishing here out of this community, and neither one of us owes a cent, not one copper" (NGP-9a, male fisher).

An adherence to an ideology of self-sufficiency and to the "traditional fisher model" is also connected to the reciprocal exchange of services among men in Newfoundland. According to Walker (1994: 48), the exchange of services among men serves to solidify friendship and offset household expenditures. Nemec (1972: 23), documenting the lives of fishers in a Newfoundland village, suggests that economic need for such services is in fact a precondition for many male-male friendships. Felt and Sinclair (1992: 46), however, make the point that, rather than a real material necessity, the provision of such services in contemporary Newfoundland reflects a lack of qualified trades people, the cultural value ascribed to such services and activities, and a sense of obligation.

No doubt material necessity, a lack of skilled trades people, cultural value, and friendship ties have explanatory value for the motivation behind the exchange of informal economic services among fishers. However, these activities also present men with opportunities to demonstrate masculinity. In other words, the culturally valued abilities and skills required to perform many of these services are distinctly masculine and, thus, participation in such informal economic activities represents another way of creating difference between men and women, especially in terms of the construction of a sexual division of labour. In addition, these activities present alternative sites for the construction of masculinity for those men who have been denied access to real resources – cultural and material. This is similar to the argument made by Connell (1995: ch. 4) that masculinity is created in alternative realms where men lack access to resources. In her work on displaced miners in Britain, Morris (1991: 174) has argued that local men use the informal economy as an arena for male competition. These arguments have partial explanatory value for the work at hand. No doubt the cultural value ascribed to such services means that men fishers are able to enhance local reputations through participation, and also can construct meaningful self-evaluations through this work. For fishers, the work has value in and of itself and connects them to their past and the "traditional male fisher" model.

Many of the men who worked in processing plants also subscribed to similar ideas regarding time and culturally valued male activities. Male plant workers who worked in seasonal jobs (particularly those men employed at the seasonally operated plant, as well as those men who held seasonal jobs at the offshore plant that operated year-round) often interpreted their time unemployed as an opportunity to perform such valued subsistence activities as cutting wood for fuel and helping family members construct new homes. One male plant worker stated: "It was seasonal here. The nature of the beast was that you could fish in the fall months in the inshore . . . . We be in the woods cutting wood, I got a furnace and that kind of stuff. And the fall of the year you be in at the berries and at the rabbits or whatever you could do, you know, to pass the time" (NGP-10). This man talked about time and work by referring to fishing, neglecting references to plant work. Male plant workers have been able to "be fishermen," to some extent, by making use of downtime, reflecting and reinforcing an attachment to self-reliance and sufficiency. At the same time, most recognized their ability to do this as being tied to unemployment insurance benefits. Some men opted for seasonal jobs at the year-round offshore fish plant. The combination of unemployment benefits and male informal economic activities made seasonal work preferable to and more economically viable than working year-round at the offshore plant. When asked if he worked year-round or seasonally, one male plant worker explained that it was not worthwhile to work year-round: "No, I come home because I would do better on my unemployment because you wouldn't get so many hours in the summer, right?" (NGP-11).

"Being a fisherman" and the "traditional male fisher" model are ideological constructs reflecting local cultural values and a version of history that celebrates the *fisherman* as the ideal. To varying degrees local men take up or reject this model, which justifies resistance to state adjustment but also reflects the structural position of inshore fishers in the industry and in wider society, as well as the material possibilities and constraints they face. It also reflects and is made possible through women's paid and unpaid work – a point I will return to in Chapter 6.

### "A COMPLETE WASTE OF MONEY"

Refusal to be removed from fisheries work and outport communities translated into a local rejection of state-sponsored retraining and educational programs, which presented fisheries workers with opportunities to train for non-fishery-related work. Despite the almost mandatory recounting of a couple of local "success stories" regarding retraining and educational opportunities, fisheries workers underscored a number of problems with the content and facilitation of such programs. Perhaps the two most common criticisms lodged by fisheries workers pertain to the cost and the benefactors of these programs. Fisheries workers complained that educational programs, like "Improving Our Odds," an employment counselling program for displaced fisheries workers, took money away from these people who were economically vulnerable and gave it to retired teachers: "This 'Improving Your Odds' was a complete waste of money. . . . They hired seventeen retired teachers to do this 'Improving Your Odds,' to go and sit like we're doing here now, right? Only we played darts or played cards or something like this, and somebody try to tell me that I can't facilitate this, right?" (NGP-17, female plant worker). Most respondents agreed that displaced fisheries workers or young professionals and recent university graduates should have been hired as facilitators.

While the dominant discourse views fisheries workers' resistance to these programs as indicative of laziness, respondents provided a host of other reasons. First, the initial expectation that the moratorium was a temporary solution kept workers from making radical future plans: "No, my dear, I didn't think it [the moratorium] was going to be that long. I figured we be only off a couple of years and be back again" (NGP-18, plant worker). Second, state financial compensation softened the impact of being displaced, limiting involvement in retraining. However, many fisheries workers suggested they experienced immense pressure to enrol in retraining or educational programs because they feared losing NCARP or TAGS income. A male plant worker (NGP-10) reiterated this point: "[W]e had people forty, fifty years old into computers and didn't go to school for years and had no interest in there, but they were there because they thought their TAGS were going to be cut, or they were told their TAGS were going to be cut."

This pressure was very real – originally a TAGS prerequisite included being "active," meaning each client was expected to "receive counselling" and "develop an action plan identifying goals for adjustment" (Human Resources Development Canada, 1998: 12). Except for fisheries workers taking early retirement options, fishplant workers were expected to participate in, among others, training, volunteer work, or the mobility assistance programs; fishers who wished to remain active were expected to involve themselves in professionalization efforts, especially training (ibid., 13). This "active" component was later dropped but the lack of information among fisheries workers regarding the "active" criterion is worrisome. Fears of not appearing active led many to enrol in some sort of educational program and respondents warned that such enrolment should not be read necessarily as support for the program. Rather, there was in many cases a disjuncture between ideas and behaviour.

A third reason fisheries workers resisted retraining related to age. Many respondents pointed to age-related problems with retraining, including difficulties associated with being in an educational setting after being out of school for a long time, especially for those who did not enjoy school the first time around, as well as problems experienced by those with little basic education and a fear of failing in school. This fear could have a tremendous negative effect at a time when fisheries workers are experiencing low self-esteem and other psychological tensions associated with being displaced or otherwise economically vulnerable. Many voiced concerns about the potential dangers involved when people are "forced" to retrain: "I didn't agree with anybody being forced into [retraining]. Like you had some people who were totally illiterate, okay, to be forced into going to school, for upgrading. These people couldn't cope with it. Look at the stress was put on them by doing that, right?" (NGP-17, female plant worker).

Many workers suggested that educational programs should have been aimed at the youth. In reality, however, young workers experienced a number of problems with life after retraining. Given Newfoundland's collapsed economy, retraining did not necessarily lead to finding paid work. Where retraining did lead to paid work, it often required long commutes or time away from one's family. One male plant worker (NGP-11) found work outside his community after completing a couple of trades courses. However, he quit due to location and insufficient income: "I did computer technician course from '92 to '94 and in '95 I did the heavy equipment course. I worked into MUN [Memorial University of Newfoundland] for a month. I quit that because it wasn't enough pay trying to live in there and run the house out here." This man became disillusioned with retraining since it had not led to any real improvements in his material well-being.

It was obvious from my interviews, however, that being away from one's family in pursuit of employment or training was more acceptable for men than women, though not always practical. Women's participation in retraining was severely limited because of the cultural imposition of child-care and domestic responsibilities and the lack of affordable daycare; in addition, the retraining curriculum was aimed at men. While men and women respondents reported a shared lack of interest or involvement in retraining programs, the reasons for such responses were clearly different and organized by gender –

including ideas about masculinity and femininity, divisions of labour, and structural opportunities and limitations. Men's and women's differing positions within productive relations also helped determine the extent and type of involvement in retraining and educational programs.

Perhaps the main way men fishers were able to escape enrolment in retraining or educational programs and removal from the fishery was through continued participation in the fishery, albeit a limited fishery. Based on my interview data male plant workers experienced more pressure to enrol in a course than men fishers: "A lot of plant workers took [a course in] tractor trailer driving and that, but that was only because more or less they were forced into it. . . . But the fishermen weren't like that because with the fishermen, we're still fishing lump roe . . . and that season probably last probably twelve weeks. So, you know, that's how they got away from taking a course" (NGP-2, male fisher). Fishers who did participate in educational programs appeared to limit involvement to fisheries-related courses focusing on navigation and safety, once again reinforcing their commitment to fishing. Some fishers indicated that fisheries-related courses were preferable to "go[ing] back to school just taking courses" (TEK-31, male fisher).

## "A WAY OF LIFE"

Male fishers have coped with the fisheries crisis by adapting fishing strategies, engaging in the subsistence and informal sectors of the economy, and rejecting retraining. Each of these coping strategies is directed towards resisting removal from the fishery. Why are fishers so determined to remain in an occupation that has been plagued by crises and has often provided a meagre living? Part of the reason lies in their desire for continuity amid chaos: "We've been here a long time. Bonavista's been here a long time and we're not planning on going nowhere yet" (NGP-10, male plant worker). This comment points to an oppositional strategy employed by local people of "looking to the past." A respect for and perceived relevance of the past extends beyond harvesting fish and the skills required on the water – although these are relevant as well. Many local fisheries people have compared the present situation to the state-directed resettlement program in the 1960s that removed people from the fishery and land and ocean resources necessary for survival. In fact, state policies dealing with the fisheries crisis have, in some instances, strengthened locals' commitment to their "way of life."

There is a general new-found relevance in the collective history of fisheries workers and in the lives of their forefathers and foremothers. Men and women drew on this collective history to make sense of the current crisis and to provide hope for survival. A historical stoicism breathes new life as locals face hard times and must learn to make do with less: "I'm still surviving" (NGP-8, male fisher). People remain hopeful, refusing to give up on the only life they have ever known: "But we're hoping, you know? You live in your hope and die in your spirits, I suppose. But we're hoping that the groundfishery will come back" (NGP-10, male plant worker). This hopeful spirit expressed by male and female respondents, however, should not be taken as evidence of a lack of problems within families and communities since the moratorium. Indeed, there is some evidence to suggest that problems, including alcoholism and family strain, are on the increase.

Much of the existing literature on Newfoundland (for example, Andersen and Wadel, 1972a, 1972b; Davis, 1983, 1993; Finlayson, 1994; Porter, 1993; Szala, 1978) has pointed to the significance of the fishery, the rugged physical environment, historical political neglect, and exploitative relations between merchants and fishers in the construction of a unique cultural maritime identity. Fishers, in general, have reinterpreted and romanticized their historical, cultural, and material subordination as a yarn of success and survival against the odds. This version of the past has provided the lens through which fisheries workers see and explain the world, make decisions, and face adversity. This "past" has also defined local values. In this way, hard work, stoic endurance, strength, egalitarianism, self-sufficiency, and resourcefulness are valued because they have enabled Newfoundlanders to survive. A shared history of struggle and commitment to the "traditional" fishery has led some to characterize Newfoundland, like other marine economies, as having a distinct culture (Nadel-Klein and Davis, 1988).[4] According to Davis (1983: 136-7), identification with maritime culture extends beyond the work experiences and opportunities in Newfoundland to influence housing patterns, styles of men's clothes, topics of conversation, and local language.

Respondents in this study reported a familial, historical, and material attachment to fishing and outport communities more generally. At the time of the interviews most of the respondents resided in communities where the fishery had been the economic mainstay. Interviews with men and women, fishers and plant workers alike, revealed that the fishery and maritime culture have supplied the materials for a collective sense of togetherness and belonging to something greater than the individual – even if the particular ways in which women and men were able to identify themselves as part of this community were gendered. The fishery has come to embody the values and way of life of outport Newfoundlanders, or to put it another way, the fishery depicts the "ethos" of Newfoundland (Szala, 1978: 85-90), in which male fishers see themselves as an integral part. Fishing is "a way of life, a social fishery" (NGP-7a, male fisher). Fishing is more than just a job to fishers and certainly entails more than simply the catching of fish: "It's what Newfoundland is all about" (TEK-33, male fisher). It should be noted that when respondents, like Newfoundlanders more generally, referred to the fishery in these ways they tended to be speaking about a particular fishery, namely the *historical* inshore cod fishery, not the modern offshore multi-species fishery (Finlayson, 1994: 4).

### "BEING YOUR OWN BOSS"

Literature on the male working-class experience has suggested that men regard any manual job as good as another. Willis (1977: 99-100, emphasis in original) states, "*particular* job choice does not matter too much to 'the lads'." The lads have bought into the working-class cultural assumption that no job is enjoyable and, thus, have invested in "generalized labour" to meet their needs for quick money. While it may be possible for working-class men to find ways to express a sense of self at the workplace, such expressions occur outside formal work routines. In this way, Willis (1977: 102) explained, "[l]abour power is a kind of barrier to, not an inner connection with, the demands of the world. Satisfaction is not expected in work." Fishers, on the other hand, expect and find satisfaction in their work that is unlikely to be found in other manual jobs. Fisher respondents listed a number of elements of work that provide satisfaction, including the intrinsic nature of work tasks, working with their hands, the intimate relationship with nature, and the freedom and autonomy experienced in being their own boss. The self, work, culture, and masculinity are all intimately connected for fishers, providing space for expression and validation – these are other aspects of the "traditional male fisher" model. Given the few meaningful and satisfying opportunities that capitalism would present to these men, many of whom lack formal education and training, it is no wonder they resist attempts at being removed.

For "independent" inshore fishers, fishing is a job like no other. Fishers derive satisfaction and a sense of self from their ability to exercise a certain degree of control over their work time and labour and from their experiences of independence and freedom: "The freedom, you're your own boss. That's probably the biggest thing, you know what I mean? Now you got no one looking down over you. You can go and come when you like. . . . That's part of it, and out on the water. I always loved it on the water, so, like I said, it's the freedom probably above everything" (NGP-1, fisher, age 35). This experience is largely unavailable in male working-class jobs, where labour is restricted and controlled by the clock. Kimmel and Kaufman (1994: 261) have suggested that the current so-called "masculinity crisis" is rooted, at least in part, in the erosion of workers' autonomy and control over their labour associated with capitalist work organizations. The case for men fishers, however, has been different in that, to some extent, they have been able to retain some independence in their craft despite increasing government involvement. As Porter (1993: 87) put it, fishers have retained some autonomy and control "at the end of a harshly exploitative capitalist chain."

Male fishers underscored their opportunities to experience autonomy by contrasting harvesting and processing work. Women who had worked in both the harvesting and processing sectors also said they preferred fishing to plant work. However, the satisfaction they received from fishing is not linked to their gender identity in the same way as it is for fishers. Women did not experience the same degree of autonomy on the boat due to their subordinate positions as crew. For the male fisher respondents, factory work challenged beliefs about the type of work that men should do. Some fishers drew on their previous work experience in a local fish plant to critique its degrading, monotonous, rushed, and heavily supervised work environment that contributed to feelings of entrapment: "You're down there piled up into that [factory] all day long. Someone looking down your back everyday, every hour, every month. . . . No, that wasn't cut out for me" (NGP-9a, male fisher). Another male fisher (NGP-2) provided a similar critique: "I worked in the fish plant for a few years and I didn't like that . . . I didn't feel right there. You were closed in, and it was more or less rush, you know? Hurry up and get this done. . . . Nice days you come out dinnertime and sun be shining and you know you had to go back in there and after a while it just got to me, hey."

Most fish-processing workers reiterated this sense of commodification in their status as wage labourers. The work environments of fishers and plant workers are drastically different. The formal rules, architecture, and dress code in the plant are all immediate indicators of subordination. Both men and women processing workers complained of being treated like robots and problems with supervisors. Especially in the context of fisheries declines, plant workers encountered increased supervision and less control over their work (Fishery Research Group, 1986; Neis, 1991; Power, 1997). Not surprisingly, however, most plant workers claimed to enjoy their work experiences. Critical questioning revealed that workers actively found satisfaction in the social aspects of work, from earning wages, and in doing a good job rather than in the actual job tasks, which were often monotonous and wearisome. A retired male plant manager (NGP-13a) said that plant work was enjoyable because there was a "great crowd" working at the plant. Another male plant worker (NGP-10) linked job satisfaction to a sense of "camaraderie" among the plant workers. Negative feelings and experiences were more likely to be reported by women than men plant workers reflecting the gendered and hierarchical divisions of labour.

Inshore fisher respondents also differentiated their work experience from that of fishers who skippered or crewed on company-owned boats by pointing to the latter's lack of autonomy. Men who crew on company vessels hold particularly vulnerable positions, experiencing a lack of control over their work, time, and incomes (Andersen, 1972: 132-8; Sinclair, 1988b: 159-60). Those inshore fisher respondents with previous experience working on company trawlers preferred working in the inshore small-boat or longliner fisheries to the offshore fishery or on company-owned boats because of the unfavourable power and social relations among crew, skipper, and company, as well as the different technologies employed in each sector. A male fisher (NGP-9a), who had experience fishing on company-owned vessels, as crew on a local man's longliner, and as skipper and owner of an inshore vessel, explained his preference in terms of work autonomy:

> NGP-9a: I prefer inshore fishing *in my own boat* anytime, anytime at all.
>
> NGP: What makes the difference?
>
> NGP-9a: Well, you're your own boss, and you can go when you like and you can come when you like. You got no one there telling you when to go, when to come. . . . When it's bad weather . . . you can decide if you can go out. If you thinks it's too bad, that you don't want to go out – you're not going to go. But if you're with someone else or if they thinks well, no, the hell with it, we're going, well you're going. It's either go or go ashore, we'll get someone else.

Trawler workers on company vessels also recognized their own vulnerability and lack of control, illustrated by their fears of replacement. A hierarchical work environment, ecological decline, and job scarcity exacerbated this sense of vulnerability and created an atmosphere of individualism among the workers: "Right now the company can do what they want to us. We got no choice. We can say, 'We're not fishing.' Skipper can say, 'Good enough. I'll find somebody else,' and he's got somebody else within minutes. So we got no choice. We got to take it or get off the boat. And around here, there's nothing to do. So you've got to take it" (TEK-10, male crew member on FPI vessel). Perhaps not surprisingly fishers employed by FPI were especially hesitant to discuss negative aspects of their fishing experiences or make comments that would reflect poorly on the company. For example, when asked to elaborate on his comments regarding abuses he witnessed as crew on company-owned boats, one fisher (TEK-31) replied: "I won't elaborate much on it, though." His silence may be indicative of his precarious position.

Based on the above descriptions of the work experiences of factory workers and fishers employed by processing companies, it appears that inshore fishers have indeed been able to exercise a certain amount of real control in their work. However, because most of the fisher respondents referred to here skippered their own small boats or longliners or were at least involved in partnerships, this experience of autonomy may be slightly exaggerated in this account. Crew members obviously have less autonomy, choice, and authority over decisions than owners or skippers. Of course, fish stock declines, increased state intervention, and restrictive policies regarding licensing, fishing seasons, and quotas (discussed later), as well as the control exercised by fish-processing companies at the local and, increasingly, global levels, serve to limit independent inshore fishers' experiences of autonomy. Fishers reported feeling pressure to ship fresh fish to the local plants (or the plant closest in vicinity) in order to maintain good relations in the region. Many fishers had relatives working at the local plants and felt obliged to ship there:

> FPI got the market cornered on everything anyhow. . . . They control this area, right? Try and get another buyer to come in and set up is very difficult because most all the fishermen that ship to FPI have their sons or daughters or somebody in their family working in the plant, and they feel that if they go somewhere else and the fish goes out of Bonavista, it's less work for them, right? (NGP-2, male fisher)

Shipping outside the community was, indeed, a source of arguments and fights between men at local bars: "But it makes for good relations in the community to ship it here. . . . There were a few scraps in the clubs. Nothing you can't handle. All we used to tell them was to go up and tell the management to pay us the price and we will ship our crab there. That is all there is to it" (TEK-18, male fisher). Fishers' autonomy was also limited by the provision of loans by plant owners and managers to fishers to purchase fishing gear and property. Attached to the loan was the expectation that fishers would ship to FPI. Some fishers suggested that the local processing company keeps fishers in debt "so then you can't move anywhere else" (TEK-18). Increased difficulty in getting loans for fishing gear and property in recent years has given fish-processing companies more control: "You can't get nothing from the bank, or very little, you got to go to the companies. It's more like being, going back to the times of the merchants" (NGP-1, male fisher).

Low incomes also strained fishers' autonomy, forcing them to sell their labour occasionally. Indeed, some fishers crewed on company-owned trawlers to compensate for low incomes earned in the inshore fishery or as a means to acquire funds to purchase boats and gear. For example, one male fisher (NGP-7a) worked on company-owned draggers during the winter months, when the inshore fishery finished, to offset losses in the inshore. Money earned from the offshore enabled him to continue to fish independently inshore. Fishers engaging in this strategy preferred to fish inshore but were often unable due to low earnings: "Oh, I sooner be inshore fishing, to tell you the truth" (NGP-4a, male fisher). How did these fishers make sense of sacrificing the freedom they valued so much? Fishers interpreted this behaviour as a sort of manipulation of the company or a strategy that would buy them freedom later. It was a means to an end, a sacrifice that would later pay off. Yet this strategy also points to the contradictory class position of these inshore fishers – simultaneously part-proletarianized, part-independent, as well as to the contradictions between ideas and behaviour.

## "YOU WOULDN'T BE A RICH MAN"

The seasonal nature of the fishery has tended to mean unemployment and low incomes for inshore fishers.[5] Men respondents experienced tensions between holding a deep pride in such work and occupying a marginalized position in an exploitative capitalist system. Fishers face a peculiar contradiction: they work hard and perform a job with relative autonomy for little money, especially as stock declines have made it increasingly difficult to find and catch fish efficiently, while big processing companies rake in huge profits through their labour. In addition, fishers' craft is embedded within a set of tensions regarding work status. While it is clear that fishing is men's work ideologically and numerically speaking and maintains high status at the level of maritime culture, it is also manual work and the financial and class rewards are low: "A small inshore boat like myself, we don't make no big pile of money. If we can make $20,000 a year, that's big money for a small boat in this area here. Some areas may make more. I've been averaging $15,000 to $25,000 in my lifespan of fishing" (TEK-32, male fisher). This contradiction connects fishers to "the past."

How do fishers make sense of this contradiction? Men actively search for meaning, identity, and power, in their daily experience of work and ways in which to justify their job "choice." At the expense of materiality, fisher respondents tended to emphasize the sense of freedom they found in the practice of their craft and the bodily sense of work, tools, and being close to nature. Herein lies a major difference between the working-class male experience and the experience of primary producers and "independent" craftspeople. As reported in Willis (1977: 150) and Collinson (1992: 86-9), working-class men make sense of their commodification by locating freedom in the wage packet, insofar as wages are earned under brutal (and thus masculine) working conditions, which confers household power and limited purchasing capacity. Fishers, on the other hand, claimed they are *not* in it for the money. This is not to say that they are not concerned about their ability to provide for their families. Clearly, they are. Nonetheless, interviews with fishers corroborate a point made by other researchers (Davis, 1993: 462; Porter, 1993: 93), namely, that men fishers are proud of their craft and of being primary producers despite the experiences of low incomes and unemployment that accompany the seasonal fisheries. Consider the following remarks made by a male fisher (TEK-27):

> When I decided to come back, the longliner people were starting to take off. Our government officials were saying, "There's going to be thousands of fish," and I wanted to do what I like to do. Wanted to make a living. But the only reason why I left the fishery in the first place is because there was no living there. You couldn't survive. So it looked like then, it's a place where you could go and survive, make a living, not make no big pile of money, but make a comfortable living.

Fishers subjectively balanced their disadvantageous incomes by investing in work that was culturally valued and enjoyed. What is more important than "getting rich" is "being satisfied" and "making a living" (NGP-7a, male fisher). According to one male fisher (NGP-1): "You wouldn't starve to death at it, and you wouldn't be a rich man, I wouldn't say, but you made a living. And that's the main thing, you know?" Another (NGP-2) suggested a similar commitment to being satisfied: "But basically everybody was satisfied if they came in and at the end of the day, if they had a couple thousand pounds. If the fishing was good, you know they were, they were satisfied with it." Holding male competition in check was also important: "When you're interested in something, you don't want to be the best, or you don't want to be the worst. But as long as you're average, the middle, you get along all right" (TEK-16, retired male fisher).

These comments demonstrate satisfaction with being average that may reflect a larger cultural and historical egalitarian ethos. Many authors have commented on the widespread egalitarian ethic in rural Newfoundland communities.[6] Egalitarianism, which levels differences among locals, can be read as a response to the relative powerlessness and marginality of fishers. This ethic mediated interpretations of the tensions fishers faced, informed their consciousness, and created a sort of resignation to the status quo. Alternatively, these comments may simply reflect conformity, at least at the level of talk, to informal social rules about egalitarianism for fear of chastisement at the community level. Commitment to an egalitarian ethic, real or superficial, demonstrated through rejections of greediness, may also reflect attempts by fishers to portray the self in a favourable light. At the same time, some fishers were quick to blame those who failed to make a living in the fishery by pointing to a poor work ethic: "If you goes out and don't get no fish, you got to blame yourself. If you don't put the effort in and you see the guy next to you is putting the effort in there and getting it, and you don't, well, you got to blame your own self, you know?" (NGP-1, male fisher). This approach to work carries highly individualistic tones and fails to recognize the effects of the intersection of state control, big business, and capitalism that have made it increasingly difficult to survive in the inshore fishery.

The notion of "being satisfied" with getting by can also be interpreted as localized resistance to dominant notions of the male breadwinner. The concept "breadwinner" and the reality of the cash economy have a relatively recent history in Newfoundland. Historically speaking, the bourgeois version of the breadwinner/homemaker divide has been largely irrelevant for households that depended on the work of all family members, within a gendered division of labour, and plural adaptations in order to survive. Rather than conceptualize breadwinning in terms of money earned, fisher respondents defined breadwinning in terms of the circumstances in which money is made and the employment of particular strategies. According to Willis (1977: 150), working-class men reinterpret the harsh conditions of manual work "into a heroic exercise of manly confrontation with *the task*. Difficult, uncomfortable or dangerous conditions are seen, not for themselves, but for their appropriateness to a masculine readiness and hardness" (emphasis in original). In this way, it is not the work itself or the organization of work that imparts a positive sense of self. Rather, it is the sacrifice and strength required to perform work, of which it is claimed not everyone – particularly women – is capable.

For fishers, physicality is an important way in which work becomes linked to masculine discourse and practice. Certainly, their ability to face and defeat a dangerous and unpredictable Mother Nature provides self-affirmation. However, unlike the working-class men described by Willis and others, fishers' formal work tasks, (as well as their informal survival strategies and position in productive relations) are intimately connected to the masculine self. Both culture and economic position shape how men's work is defined. The working-class definition of "breadwinner" is tied to the "manly confrontation" needed to earn wages as well as to the limited purchasing power it brings. Male fishers, on the other hand, tended to link the breadwinner concept to "being satisfied," work autonomy, and "being a fisherman." In this way, fishers' breadwinner discourse links the shortcomings of capitalism's cash economy to the need for subsistence and informal work and reflects the contradictory class position of fishers. While fishers resisted dominant definitions of breadwinning, they were very much aware of the increasing pressure to use wealth and consumer items as displays of status. Their households, like others across Newfoundland, are inundated with Canadian and American versions of the "good life" presented in the media, especially television. Fishers do not easily fit into these versions.

For fishers, freedom means more than simply being your own boss. It also refers to not being confined by the four walls of a factory. The vastness of the ocean, the feel of the saltwater spray, and a visceral relationship with the power and beauty of nature are other aspects of fishing that provide satisfaction and make it a job like no other. It is not only "being your own boss" but the fact that work takes place "outdoors" that makes fishing worthwhile (NGP-2, male fisher). Fishers interpreted their work environment as a form of self-expression. The act of fishing itself – working with one's hands and the work tasks involved – is important in the construction of masculinity because of the links to physicality, physical risk, and embodiment. Working outdoors, at sea, brings with it a number of physical risks and experiences. As Andersen and Wadel (1972b: 144) have noted, the "stereotype behaviour of the fisherman (and sailor) is characterized by extreme masculinity display: physical strength, endurance, fearlessness, and stoicism are highly valued, and virility is a prominent conversational theme." They report, however, that this display, while common among men who take part in fisheries that require prolonged voyages with large crews, is not an accurate description of the gender display of fishers who take part in day trips like most of the fishers discussed here.

While overt masculine displays may not be evident among small-boat fishers, the "dramatic intensity to fishing" nonetheless informs male subjectivities and shapes imagery used in descriptions of male confrontations with "an unpredictable and dangerous mother nature" (Davis, 1993: 462). Studies have recognized fishing as one of the most dangerous occupations (Murray et al., 1997: 292): "Well, it was dangerous work you know what I mean? . . . But we never feared nothing like that. . . . I used to go out with me father when I was perhaps six year old . . . but as we grew older that [fear] wore away from us, you know? We didn't care then, you know?" (NGP-3, retired male fisher). Recalling a particularly rough day at sea, one male fisher (TEK-33) drew attention to how the element of danger moulds fishers' psyche over time: "And I be there cringing. Really get rough up there. When I got out of school first, I wasn't cut out for a fisherman. I really wasn't. I was just more or less at it to make a few dollars and get away. And I don't know. I just start to like the danger I guess. I wouldn't trade it for the world now." Fishing entails such very real dangers as drowning and getting lost and such risks as having gear damaged and catches lost due to poor weather conditions.

Fishers deal with these sorts of unpredictable factors on a daily basis. Many likened fishing to gambling. Nemec (1972: 15) suggested that this perception that fishing is a gamble is a fair estimate of the associated ecological, economic, and technological limitations. Echoing this assessment, one male fisher (NGP-5a) commented on such constraints: "If you had trouble, had a breakdown, well I always seem to have it when the fishery was going, and you miss out on it. It's gone . . . all down the toilet, you know? . . . It's like gambling. I don't gamble. I don't play none of them slot machines . . . but I gambles enough with my life. I gambles enough to pay for it anyway, but there's a lot of gamble there." The weather often proved to be a formidable obstacle. The unpredictable and dangerous nature of fishing, though, may encourage a certain degree of co-operation among fishers. Fishers have to trust each other for safety reasons, keeping competitiveness in check to avoid potential divisiveness and danger. One man (NGP-2, fisher), who had lost a brother to the sea, discussed the importance of co-operation and safety:

> Sometimes when you get out the weather happen to turn bad, you know? The wind come up quick. It's a worry, you know? It's a worry for anybody, you know? Because I know myself, now I got a fairly large boat and sometimes when you come in you see a guy,

> eighteen-foot boat. You be trying to watch, watch the waves and that, you know? Trying to keep an eye on him too. Make sure that when you gets in he's coming right behind ya or something like that, right? Most everybody look out for one another.

In addition to the obvious dangers associated with working at sea, fishing also requires a certain stamina to cope with physical work. Fishers reported looking for "good, stouty boys" who were able to withstand the work as crew (TEK-12, retired male fisher). Indeed, such physical prerequisites have been used to exclude women from fishing and to legitimate such exclusion. However, in the views of some fisher respondents, not all men could handle the hard work and long hours either:

> . . . I'm after having a fair share few with me, you know? Comes and goes. A lot of people don't stay. Hard work fishing, long hours. . . . I likes all hands working. . . . Don't like few fellers having to pick up for someone else's slack, you know? That's no good. He's getting the same as he's getting, so he should work just as hard as he's working. I knows, I don't call it hard work, just long hours that's all, long sleepless hours. (NGP-5a, male fisher)

Fishers took pride in the physical work, in their ability to cope with the rigours and danger and in having a reputation to do so. A man's concern about being a reputable fisher had the effect of reinforcing a commitment to physical work. According to Felt (1987: 22), pressure to conform characterized men's experiences in rural Newfoundland. Local definitions of "hard-working" and "lazy," in this way, may be considered forms of social control: "If you never went out in mid-May you was lazy around here. A fellow would make fun of you" (TEK-44, male fisher).

A paradox experienced by these men is that fishing both reflects a lack of choice in a harsh environment and is a creative embodiment of who they are and the values of outport life. The construction of the self in this way reflects their real-life material conditions and relative positions of power. In addition, fishers' investment in an occupational group that equated fishing work with "real work" served to divide men and women, men fishers and other local men, as well as men fishers and men with formal power and education, including government officials and scientists (as I discuss later).

### "EASY MONEY"

On a local level, it might make sense to conceptualize masculinity constructed around the beliefs, ideas, and practices of men fishers to be hegemonic, implying some sort of hierarchy of masculinities within fishing culture. Like Porter's work on Aquaforte (1993: 87), I found that many of the male plant workers accepted fishing as the ideal, expressing a desire to enter or return to fishing when it was economically viable for them to do so, or at least, positioning themselves as closely to harvesting as possible. For many, participation in non-harvesting work had become a means to an end because inshore fishing did not provide adequate income. Given the local value ascribed to fishing, what does this mean for men who work in fish plants? What did men processing workers lose and gain through this work?

As described above, it is clear that men (and women) do lose a certain amount of autonomy by working in fish factories, where their time, work, and quality of work are controlled and monitored. However, men plant workers tended to interpret their wages as a source of economic power. This parallels the interpretations of other working-class men (Collinson, 1992; Willis, 1977). Women plant workers, equally, regarded their jobs as a means to the end of getting wages: "I mean, you go in and do your day's work and get your day's pay and that's all we were concerned about, right?" (NGP-18 female plant worker). However, the labour process and experience of waged work were different for women and men because of the sexual divisions of labour in the plant and in the home.

Men respondents who worked in fish plants offered a number of reasons as to why they entered the processing sector. Plant work provided men who lacked access to inherited fishing property, or who were simply not interested in fishing, with an opportunity to do fisheries-related work: "A lot of my, like my uncles and that, they were in the fishing boats. I just, I happened to get a job in the plant. I wasn't interested in going on the water" (NGP-10, male plant worker). Like fishing (see Chapter 4), acquiring fish-plant work often appeared accidental. Some men did not choose plant work. Rather, they fell into it at a time when the plant offered plenty of work opportunities. Others gained entry through their fathers. In addition, local communities offered few other opportunities for paid employment and men respondents often lacked formal qualifications required for professional jobs available in or outside local communities. Boys could simply "walk out of school and go to work and make good money" (NGP-14a, male fish plant worker).

A major difference between men processing workers and fishers is the extent to which the former invested in the wage, which in this case was non-professional but relatively decent,[7] to justify their job choice; the latter invested in culturally valued work. Plant work allowed people to remain in their own communities and live comfortably: "We were in our own community. Wages were pretty good and we used to work good seasonally and we get good UI [unemployment insurance], right?" (NGP-10, male plant worker). However, the wage proved to be a weak symbol of masculine practice and identity for male plant workers due to seasonal unemployment, changes in the organization of work, adoption of labour-saving technologies, and ecological declines that threatened redundancy and job loss.

While some fishers complained about earning less income than processing workers, the "easy money" (NGP-7a, male fisher) available at the fish plant contradicted their ideas about what constituted meaningful men's work. Fishers simultaneously recognized and rejected the possibility of individually swapping their positions as primary producers for positions as wage earners at the plant. Some fishers felt both resentful and betrayed, because in their view the plant workers were "making the biggest kind of wages" (NGP-7a, male fisher) for fewer hours. One fisher (NGP-7a) went as far as to call plant workers, "little millionaires," citing various displays of wealth including the purchase of big consumer items such as camping trailers as evidence. Particularly problematic for these fishers was their belief that plant workers acquired this material wealth on the backs of fishers' hard work. A retired male fisher (TEK-17) linked the introduction of fish plants in terms of incomes and the difficulties that fishers faced in recruiting male crew:

> . . . you wouldn't make near as much as they were in the plant. And you couldn't keep, men wouldn't go at it for that. The fella who owned the boat, he might be just as well off. . . . It was hard to get them [sharemen] then, when the plant got there. Hard to get good men then. . . . When the fish plant got there, you couldn't get enough fish to make a day's pay. The ones working in the plant were doing better than the ones that were fishing. They'd be going out six o'clock in the evening dressed off and you would be going down to the stage to bait up your trawl.

Such evaluations have left fishers and plant workers divided. However, the gendered divisions of labour in processing factories and in the home have also shaped the construction of masculine discourse and practice of men plant workers. The fact that men's processing work took place within a gendered division of labour, which delegates to men the "tough" or "more skilled" jobs that women supposedly cannot do, supported particular constructions of masculine practice and subjectivities. This division of labour encouraged men plant workers to see factory work as a site where men could be men. Willis (1977: 150) has pointed out that even where wages are low, their connections with the breadwinner role hold fast among working-class men because it is "won in a masculine mode in confrontation with the 'real' world which is too tough for the woman." Even where a woman works and her wages are necessary for the survival of the household, this gendered division of labour ascribes a particular value to men's work not ascribed to women's paid and unpaid work. In this way, the gendered division of labour in the plant helps explain that while husbands and wives may both work at the local fish plant, the husband maintains the breadwinner position.

The ways in which local men constructed a sense of the masculine self and invested in masculine practices were mediated by their economic and social positions, linked "to the real-life possibilities of these men and the tools at their disposal for the exercise of some form of power" (Kaufman, 1994: 145). Unable to really "be fishermen," plant workers drew on other sources to make sense of their work and of themselves. Plant workers used the wage and the gendered division of labour at the plant as their props to demonstrate masculinity; fishers invested in more abstract ideas about struggle and survival and commitment to the "traditional male fisher" model, as well as the gendered division of labour in the fishery. While the props may differ, the process here is the same. How local men construct a sense of themselves as men is, as Kimmel (1994:125) puts it, "about the differential access that different types of men have to those cultural resources that confer manhood and about how each of these groups then develop their own modifications to preserve and claim their manhood."

Because fishing – as defined by the "traditional male fisher" model – is interpreted as more than just work and represents a form of resistance to dominant definitions of progress and the modern work experience, it is not surprising that men fishers resisted removal from the fishery. Fishers' resistance was based on the observation that besides fishing work, there is little meaningful work available in their communities. Men fishers' subjective and economic investment in their work, combined with the cultural respect assigned to fishing work, invalidated a search for other types of available work that fail to provide personal satisfaction. Fishers' rejection of retraining and early retirement options, and their refusal to sell back fishing licences to the government and to move from their communities, reflected active resistance to current policies aimed at reducing the number of inshore fishers and, consequently, collective power. In fact, inshore fishers are in a unique position relative to other fisheries workers, who depend (at least more directly) on an employer to hold on to their way of life and avoid retraining and moving from their communities. Fishers have interpreted the state's imposition of moratoriums as an overt display of power that directly challenges the very aspects of fishing work in which these men have invested. Men fishers have cultural, economic, and subjective investments in fishing work. It makes sense, therefore, that men fishers reported feeling disturbed by the possibility of the need to acquire degrading work at some point to earn money.

Yet inshore fishers, the focus of this book, find themselves in a very different position within productive relations, and they work under different circumstances from those of "other" working-class men. Fish-processing workers, on the other hand, are in a similar work position to that of the factory workers described by Collinson and Willis. The particular set of circumstances in which fishers are located, no doubt, impacts the construction of their identities, subjectivities, and practices as men. Inshore fishers are part of the primary sector. Most own and operate their own fishing vessels, either alone or in partnerships, or serve as crew on a vessel owned and operated by a family member or acquaintance. In contrast to the working-class experiences of those who exchange their labour for a wage from an employer, these fishers do not work by the clock or experience direct supervision. Rather, fishers have a fair amount of autonomy. Such experiences of work have explanatory value for fisher respondents' interpretations of their work as a vocation rather than simply a job. Fisher respondents, thus, did not separate their work from their sense of self, as working-class men do. At the same time, fishing work has a number of things in common with other manual work, such as insecurity of income and dangerous working conditions. These conditions, combined with increased government intervention and dependence on powerful processing companies, serve to limit autonomy at work and outside.

A major theme running through the interviews with fishers from the Bonavista Bay region is that small-boat fishers were not freely able to sell their catches competitively. They described two main alternatives – they could sell their raw product either to one of the local FPI plants or to independent, small processing companies located outside the local community. Most fishers opted for the former for three main reasons. First, inshore fishers reported a good working relationship with management at the Bonavista processing factory, primarily because inshore fishers supplied the plant. Second, fishers experienced pressure to ship fish locally rather than to independent plants located outside the community for a variety of reasons, as discussed earlier. Third, fishers reported that dealing with smaller companies, especially during ecological decline, entailed a certain amount of financial risk. FPI, on the other hand, was regarded as reliable, especially in terms of payment, because of its size: "A lot of them small companies all promise you the moon, 'We'll give you this, we'll give you that.' The way I got FPI sized up, they don't come along promising you this and promising you that. If you're with 'em, you're with 'em, and if you're not, you're not. They don't care" (NGP-5a, male fisher). Despite this element of risk and unreliability, some small-boat fishers decided to sell their fish to independent plants because they offered better prices and treated fishers with "respect." However, the moratorium has devastated many smaller independent fish plants, compelling more and more fishers to deal with larger companies like FPI.

The limits to inshore fishers' autonomy and their reliance on processing companies have been made quite explicit on occasion. A number of male fisher respondents complained about the experience of exclusion and marginalization in dealings with FPI's Port Union plant. According to these fishers, management at the Port Union plant did not care about the "little bit of fish that inshore fishers brought in" (NGP-7a) because company-owned offshore fishing vessels secured a constant supply and "this plant [the Port Union plant] was up here was open but he didn't want nothing to do with the inshore" (NGP-5a). Another accused management at the Port Union plant of treating him and other small-boat fishers like "second-class citizens" (NGP-7a). This phrase suggests that fishers perceived an imposed differential status system between "us" – the inshore fishers – and "them" – the "modern" offshore sector. Certainly, this perception reflects real power relations and FPI's ability to direct local fishers. Many fishers, however, failed to see this as a corporate problem and reduced the situation to the individual personalities of the managers at each of the plants.

According to some fishers, when FPI docked their trawlers in the early 1990s, though, the tables have turned somewhat:

> I never had no time for them [management at the Port Union plant]. No, I'll tell you why. Because back years ago they did us awful wrong, not only us but they did all the fishermen awful wrong, right? When the trawlers was there, the longliners and the small boats they didn't want to have nothing to do with 'em, you know? They had their fish coming in and this was all was to it. It was only a bother for them for them to go and bother with the small boats and the longliners, right? But now when the trawlers gave up, now it's throwed right in their face now . . . they were crying for us, right? (NGP-9a)

It appears that the moratorium on cod may have temporarily altered the relations between fishers and processing companies, until, that is, FPI underwent massive restructuring. Make no mistake, FPI – like other multinational companies – has tremendous control over the lives of fishers. This was clearly demonstrated in the decision made by processing companies in the summer of 2003 to discontinue buying and processing crab from fishers.

## SUMMARY

Small-boat male fishers have largely resisted, both ideologically and in practice, state-directed efforts to downsize the fishery. They have engaged in a number of coping strategies that reflect culturally specific gender configurations of practice and ideas. "Looking to the past" and drawing on the "traditional male fisher" model – however partial, contradictory, and simplified these may be – provide tools with which to respond to potentially devastating changes in the fishery and in their lives, and through such response they have been able to develop, at least to some degree, a framework to make sense of the changes. This framework, which reflects fishers' position in production relations, helps organize possible responses to perceived and real threats of displacement and disruptive change. Fishers coped by adapting usual fishing strategies, engaging in illegal or informal economic activity, intensifying subsistence work, and rejecting opportunities to train for non-fisheries related work. Individual fishers made decisions regarding which strategies to take up and to what extent. These strategies reflect structured opportunities available to *men* and to *fishers*, as well as constraints; they also reflect male agency and the importance of the local in mediating the effects of restructuring.

Men resisted downsizing and efforts to remove them from the inshore, small-boat fishery in an attempt to maintain some sort of continuity in their lives. Fishers drew on lessons learned from "the collective past" to get by. In fact, not resisting was largely unthinkable as fishing is "a way of life," and there are few, if any, real alternatives to fisheries work in rural outports. In this context, fishing is deemed the best possible option, allowing fishers to experience the rewards of performing culturally valued, "masculine" work and a certain degree of autonomy in their day-to-day working lives. At the same time, small-boat fishers have tended to earn low incomes, periodically necessitating their shifting to other sources of employment, and they depend heavily on local processors to buy their product. It is in this semi-proletarianized position that we find potential for collective action and divisiveness, opposition and acceptance of capitalist integration. While fishers experience many of the vulnerabilities of the working class, they do not always identify as proletariat. In fact, fisher respondents in this study used plant work as a way to differentiate between themselves and other men. Simultaneously, the "traditional male fisher" model is subversive in its conception and evaluation of time, men's work, and breadwinning, yet reactionary in its assumption about its maleness. Despite the widespread adoption of similar strategies, the ways that men coped tended to be individualistic and family-oriented. Cultural constructs, including the notions of "self-sufficiency" and "being satisfied," support the neo-liberal agenda. At the same time, inshore fishers' marginal position in relation to capital enables the development of a pointed critique of government-led fisheries management and science.

# *Who's To Blame?* 4

## FISHERS EXTERNALIZE "THE SOURCE OF THEIR MISERY"

Male fisher respondents interpreted the fisheries crisis and restructuring in varied ways, ranging from seeing "progress" and its consequences as inevitable and beyond their control, to developing a critique of rationalization and modernization, as well as its technological forms. In the period immediately following the declaration of moratorium a lot of public and private time, effort, and money was spent assessing just who or what was to blame for the most recent crisis. Fishers in this study tended to allocate blame outside the individual, family, and community and onto the state, processing companies, and fisheries science. The perception of the continued relevance of the past – the "looking to the past" perspective – and the "traditional male fisher" model provide a conceptual framework within which fishers make sense of their world. The tumultuous "modernization" scheme imposed on the household-based inshore fishery – including resettlement and cash inputs into processing and harvesting technologies – has become fodder for a critical assessment of the role of the state in the development and demise of the Newfoundland fishery. (For arguments regarding the negative impact of modernization and rationalization policies on the inshore fishery, see Antler and Faris, 1979; Nemec, 1972; Wadel, 1969.)

This version of history has enabled fishers to develop a means of "externalizing the source of their misery" (Margold, 1994: 28) in their attempts to know the causes of stock collapses. Given the fact that the inshore cod fishery has formed the basis for Newfoundland traditions and culture for hundreds of years, it is not surprising that fishers interpreted the state-imposed ban on fishing cod, which restricted fishers' ability to work and earn a living, as a direct attack on their "way of life." This is not to suggest that, given the decline in fish stocks, fishers disagreed with the necessity of moratoriums. However, they were highly critical of the processes that led to such drastic measures, for these processes, in their view, were not of their making. Fishers blamed the current fisheries crisis on government mismanagement, a greed-driven industry, and the processes of formal science, which they argue have destroyed the ecological base and the sustainability of the inshore fishery. A history of marginalization and fishers' respect for this history provide the materials and perceptual framework supporting their suspicions that the state has sold out the inshore fishery to the interests of big business, and this version of history has enabled the development of a critique of a heavily regulated fishery.

Fishers' arguments about mismanagement of the fishery say as much about who they are and who they want to be as they do about their "oppressors." Their arguments also demonstrate a recognition that the local hegemonic version of masculinity – constructed around the beliefs and practices of men fishers, especially the "traditional male fisher" – is in contrast to the version of masculinity extolled by the wider culture. According to this latter version, the ideal *fisherman* is fully modern, meaning that he is rational and educated, uses efficient technologies, and is integrated into the capitalist system of production. This "modern male fisher" model, however, is increasingly gaining strength at the local level as revealed in the contradictory discourses presented in the next chapter. Nevertheless, using the "traditional male fisher" model as a conceptual framework, fishers were able to make connections between the wider structures that facilitate this modern version and the current ecological crisis. According to fisher respondents, politicians, bureaucrats, and scientists have mismanaged fish stocks because they lack an understanding of the fishery, which can only be acquired through the day-to-day practice of harvesting. It is through the experiential, bodily, and patrilineal learning processes – another aspect of the "traditional male fisher" model – that fishers come to know the "truth" about their work environment and the real impact of state and business directives that marginalize the traditional fishery and its participants. A widespread local belief is that this "truth" is unavailable to those outside the male fisher position. Even women fishers were denied access to such truth, as they could not claim to have the same lifelong experiential training as men, nor could they, in the same fashion, claim to have saltwater "in the blood." The crux of my argument is that male fishers' insights, and their local validation, reflect the position of men inshore fishers in a complex set of class and gender relations and also serve to reinforce these relations. The insights are important as they offer clues to the requisites for sustainability and for making sense of relations among men and between men and women.

## BLAMING THE STATE AND THE "SUITS" IN POWER

The introduction of formalized licensing, quota systems, and fishing seasons in the post-Confederation period has transformed the Newfoundland fishery from a largely unregulated to a highly regulated industry (Department of Fisheries and Oceans, 2001: 8-10). Control over access to fisheries resources has, thus, moved out of the hands of fishers and into the hands of state politicians and bureaucrats (Sinclair, 1988b: 157, 163-70). This shift in control exposed a number of assumptions dominating policy following confederation, including the belief that the future fishery needed educated and "modern" *men* – women would hold more temporary and limited positions – to participate in and run this new rational, vertically integrated fishery (Wright, 1995: 141) and that the local, experiential, and generational knowledge and skills contained within traditional fishing households were "backward" and impeded progress.

Fisher respondents indicated that this shift in control has been detrimental to the fish stocks and to the sustainability of the inshore fisheries, largely because the "suits in Ottawa" (NGP-7a, male fisher) and "those big fellows like Brian Tobin"[1] (TEK-3, male fisher) have little understanding of the social and ecological reality of the fishery in Newfoundland and manage the fishery from a central Canadian perspective, placing value on a modern offshore fishery at the expense of a "traditional" inshore fishery. Fishers across generations expressed this anti-Canadian or anti-mainland sentiment: "They've [Canadians] robbed Newfoundlanders" (NGP-3, retired fisher, age 71). Although fishers judged particular individual politicians harshly, they tended to recognize that the problem was larger than any individual. Rather, the problem could be found in social processes. One male fisher, not unusually, readily placed the blame for fisheries declines on federal and provincial politicians who have somehow lost their "common sense" through education and are therefore different from fishers. The implied message, of course, is that he and perhaps other fishers have such "common sense":

> How we got where we are is basically through politicians and bureaucrats. The fishery runs from Ottawa. The fishery was run by people who didn't understand the fishery. Bureaucrats and politicians in Ottawa, the majority are educated beyond their intelligence.

> . . . They don't know what the Atlantic Ocean is, haven't got a clue. That's basically the problem. . . . It wasn't managed to benefit us, since '77, and prior to '77 it was a free-for-all. . . . They haven't got a clue. Brian Tobin, for a Newfoundlander, coming from the West Coast, with the education that he's got, he's got no common sense. . . . If there was a miracle, and the fish came back, this fall there was thousands of fish, give them five years and they will be in the same spot because the same crowd with the same mentality in Ottawa is going to run it. And they're going to run it from a central Canadian perspective. (TEK-11, male fisher)

Fisher respondents expressed an overwhelming sense of unfairness about and distrust of government-imposed fisheries regulations. Fishers cited numerous examples "proving" that the directives of the state and its officials are to blame for the current ecological crisis and the marginalization of the inshore fishery. In particular, fishers were critical of rules, and the fisheries officers who enforce them, that lack an understanding of the practice of fishing and the ecological implications of particular harvesting practices. One male fisher, for example, critiqued existing regulations pertaining to fish quotas and bycatch as ecologically unsound and illogical: "They say you can't bring it [bycatch] in, but dead fish is dead fish. I would never heave it away. I would sooner come in and heave it up on the wharf and let every man, woman, and child in the community come and get a meal, rather than heave it overboard" (TEK-31). Other male fishers argued that fisheries regulations that compel fishers to take aboard their boats the volume caught in their fishing gear by penalizing them for dumping fish or giving excess to other harvesters encouraged unsafe practices: "[B]ut what if there is more than you can carry? What are you supposed to do with it? If there's a vessel alongside, I cannot see the reason why you cannot give it to that vessel. . . . It's unsafe too. It could be wind blowing" (TEK-46, male fisher). Fishers' distrust towards the state and its representatives and the ways in which they place blame reflect the very real effects of historical and social marginalization. At the same time, these are strategies designed to deflect responsibility away from themselves.

### "I GOT TO MAKE A DOLLAR"

Most fisher respondents were acutely aware of their role, however small, in ecological destruction. Yet they tended to frame their contributions to such destruction in terms of need. The "traditional male fisher" model challenged aspects of the male breadwinner role; however, in the context of an industry embedded in capitalist relations, fishers could not escape it. Fishers suggested that before the cod moratorium household material need compelled them to fish cod and other declining species despite the impact on the environment or their concerns regarding sustainability: "I got to make a dollar. I got to survive" (NGP-5a, male fisher). Declining stocks caused some fishers to respond by reducing the mesh size of fishnets (acknowledged as an unsustainable practice in that it harvests young classes of fish before reproductive maturity), as well as increasing investment and effort, in circular fashion (Neis, 1992). In addition, DFO policies that disadvantage fishers who do not utilize licences (Fish, Food and Allied Workers, 1994a: 8) and the economic imperative to enter as many fisheries as possible, especially since the moratoriums on a variety of groundfish species, have required fishers to continue participating in declining and "wasteful" fisheries such as lump and capelin,[2] despite ecological concerns: "I mean you got to try to qualify for unemployment. If I don't go at that [fishing lump] I get less unemployment, right? No, if it was up to me I wouldn't put a net out, to get the lumps. They're getting scarce too, right?" (NGP-8, male fisher). Interpreting individual protest as being futile, serving only to induce financial hardship, fishers reported feeling forced to follow the lead of others and continue fishing in order to make a living, even where this behaviour directly contradicted their beliefs: "I don't believe in the capelin fishery, never did. I fished it, but, if everybody else is at it, there's a dollar to be made, I might as well do it, too. But I was a total believer that the capelin fishery should never be touched" (TEK-31, male fisher).

In general, fisher respondents denied being responsible for transforming or having the power to challenge the structural problems that they identified in the fishing industry. This denial reduced their oppositional stance to a perhaps reluctant and forced acceptance of the status quo. In their view, it is the government's responsibility to protect fish stocks and fishery-dependent families. As such, fishing practices reflect, in part, state directives. How could one blame fishers, they argued, for the adoption of smaller and unsustainable meshes for harvesting fish. After all, the state, not fishers, has the power to set such requirements: "To get back to the government, who could you blame? Could you blame the fishermen or blame the government? The government had full charge of having the twine come from other countries, and they brought over three-inch, three-and-a-half. You were legal to buy it and whose fault was it?" (TEK-34, retired fisher). Many felt that as long as the government keeps a fishery open, people will fish: "They're getting at it because everybody else is going at it. There's a fellow coming down with a load of capelin and you're tied up to the wharf. So that encouraged everybody about it then. . . . If they [the government] closes it, well, nobody won't go at it. But the fishery has got a lot to do with the government" (TEK-45, male fisher). Simultaneously, many fishers expressed a sense of regret for not being more vocal over issues concerning the health of stocks and a sense of impotency regarding control over political efforts. In addition to economic pressures and the nature of DFO policies, fishers listed geography as an impediment to collective action. In a large country such as Canada, it would be a mistake not to recognize the isolating and marginalizing effect of being located on the geographic periphery. The centre for federal politics in Ottawa and even provincial government departments were considered remote from the lives of these men.

A sort of fatalism characterizes fishers' ideas about the prosecution of the fishery, a sense that things will never change and that fishers cannot make a difference in the management of the fishery. When asked what he thought fishers could do to help manage the fishery sustainably, a retired male fisher (TEK-51) responded: "I don't want his job [the Minister of Fisheries]. I know he's making a lot of money. He can have it. I don't know what you could do to get it back, if there's anything." Another male fisher (TEK-53) replied: "I don't know if we can do much only complain, write letters." Interviews indicate a general feeling of distrust of the state's apparently new-found willingness to include fishers in the management regime: "Fishermen should have more say in the management of the fishery, if they are heard and listened to. But what's the good for you to have some say or for me to have some say, and when it's passed on to Brian Tobin and all those fellas, they're going to make their own judgement on it, anyway? What can you do about it?" (TEK-16, retired male fisher).

Denying responsibility highlights the perceived lack of power in the lives of fishers and perpetuates fishers' perception of being different from politicians. At the same time, however, blame was not always oppositional to authority in nature. Some men blamed other fishers, who contributed to such unsustainable phenomena as ghost nets[3] and reduced the mesh size of fishnets in order to catch smaller, younger fish: "I don't really blame it [fisheries declines] on DFO. I blames a lot of it on fisherman. They put the pressure on the government to change the mesh size. 'We can't catch no fish in that.' And then they'll change the size, and then everybody is getting lots of fish again" (TEK-48, male fisher). Blaming others has the effect of portraying oneself in a favourable light. Deflecting responsibility in this way reflects both opposition and accommodation to existing problems in the fishery.

"THEY ARE THE GOVERNMENT"

Men fishers were especially critical of the relationship between the "suits" and large processors. The two processing plants (one supplied by inshore fishers, the other by offshore boats) that employed the workers in this study were owned and operated by the multinational company, Fishery Products International. As fishers saw it, processing companies had gained too much power in dictating the direction of the fishery. According to fishers, state policies reflected the interests of big business at the expense of sustainability of the resource and the inshore fishery. Many fishers were quick to draw attention to the state's history of backing processing companies and the implications of such intervention. For example, the state intervened with financial help when processing companies were threatened by bankruptcy in the 1980s and the subsequent restructuring of the processing sector resulted in the creation of two super-companies, Fishery Products International and National Sea Products Ltd.(the former being more dominant in Newfoundland than the latter).

Sinclair (1988b: 174-5) argues that the interests of fish-processing companies have been well represented in government and fisheries policies. According to Sinclair, "Federal fisheries policy has been explicitly directed towards assuring the economic viability of existing fishing enterprises, especially the largest processors." He also reports that the provincial government has followed the federal government's lead in this case. Some fishers voiced concerns about the impact on their autonomy of the efforts of the governments and companies to increase the penetration of capital and further subsume labour. Some even went so far as to say that companies control the government rather than the other way around. Equally, many fishers were unable to see any alternatives to state-directed fisheries management and were resigned to an uncomfortable acceptance of the relationship between processing companies and the state:

> If Vic Young[4] tells Brian Tobin tomorrow, "Brian, I think we'll have two days fishing for the people," the next day she'll be open. Brian Tobin don't listen to us [fishers] if we say, "Brian, I think we should have a day or so." Vic Young and a few like they say open the cod fishery for a week, it would be open, but if they don't say, "Open it," it won't be open. (TEK-12, retired male fisher)

Resignation to this reality, of course, acts as an obstruction to collective action and perpetuates the status quo. Consider the words of one retired male fisher: "You might not think it, because Fishery Products and National Sea is the government. Who else is they? . . . Whatever Vic Young's got to say on it, that's it. He's the biggest push that's into it. . . . *That won't be changed* because there's too much money in them plants to change it" (TEK-16, emphasis my own). Nor did most fishers believe that the fisheries workers' union, the Fish, Food and Allied Workers (FFAW), would have any impact in efforts to change the relationship between the state and the processing companies: "Vic Young is going to have more say than what the union got" (TEK-11, male fisher).

The profit ethic of multinational processing companies comes into contrast with the "being satisfied" ethic of the "traditional male fisher" model. Many fishers argued that this profit ethic helped institute unsustainable work practices and technologies. Processing companies were a target for fishers because of the methods employed by corporate-owned offshore trawlers to harvest fish. Fishers perceived these methods to be environmentally unfriendly. Fishers who crewed on company boats either in the past or at the time of the interviews addressed, sometimes hesitantly, issues of ecological destruction and the irrationalities of the offshore fishery. One male fisher (NGP-9a) drew on his first-hand experience to speak about the practice of dumping fish by company-owned offshore boats:

> Well as far as I'm concerned that [offshore boats] should have been took off the water twenty years ago because trawlers, because it's unreal. If you seen the things that I seen on the trawlers, my dear, it'd frighten you . . . I guarantee you. Out of the sixteen years I was there, you bring up twenty-five or thirty thousand, maybe forty or fifty thousand pounds of fish and see half of it going down over the wharf . . . floating away. It's unreal . . . it wasn't reported, you knows it wasn't.

The destructive and wasteful relationship with nature characterizing the modern fishery, the increased reliance on technologies extending the fishing season nearly year-round by circumventing fish migratory patterns, and the construction of processing plants that needed year-round production to be viable were major concerns for these men: "[L]ook at the plant in Catalina, super-plant.[5] They needed to process there twelve months of the year to keep it going. And there's no way, there's no way that . . . it could last" (NGP-2, male fisher). This modern fishery was contrasted, especially by older fishers, to the conservationist "traditional" inshore and salt fisheries: "Everything is all modern now and it is no good because the fishery is gone. And that's what done it. So much modern equipment" (TEK-14, male fisher).

Neis (1992: 159) reminds us that while traditional fishing and processing practices had a conservationist effect, this might reflect technological, spatial, and labour limitations rather than a conscious approach to the environment. Some fishers echoed such evaluations: "I think today, if people were still using the typical old Newfoundland trap, you wouldn't be destroying any fish. The fish got a chance to go around the trap and they got big mesh to get out through" (TEK-26, male fisher). For some inshore fishers, the solution to some of the existing ecological problems in today's fishery lies in returning to the "old ways." They resisted dominant evaluations that defined the traditional inshore fishery as "backward." Instead, they sung the praises of this "traditional" fishery and elevated traditional methods over modern methods. While their critiques of offshore methods and corporate harvesting and their proposed solutions reflect very real ecological concerns, they also enhance the importance of traditional fishing practices and, by implication, of themselves. In this way, arguments about the greater sustainability and skills associated with inshore, traditional fishing as opposed to the offshore fishery are also about establishing a sense of self-worth. Moreover, these valuations represent an attempt to secure their place in the precarious future fishery. Inshore fisher respondents argued that because they harvest fish, for example, using fixed gear and waiting for the fish to come to them, as opposed to the hunting methods employed in the offshore fishery, the inshore fishery is more sustainable. Nonetheless, new fish-finding technologies adopted by inshore fishers challenge this view. At the same time, and perhaps for the same effect, some fishers accused colleagues of engaging in unsustainable practices like releasing harvesting gear, usually subsidized, in the water, which then continues to gather fish for as long as the nets remain intact: "Back a few years ago, the fish companies were supplying a fisherman with all kinds of gear . . . and most of them put the nets out, and they left them out. But anybody who was paying their way didn't leave their nets out" (TEK-22, male fisher).

Fishers also tended to argue that small-boat fishing is more conducive to quality fishing, meaning fishing gears and technologies employed do not damage what is caught, whereas the methods employed in the offshore fishery sacrifice quality for quantity. They remarked on such fish removal methods used by company trawlers as suction, which affected the texture of fish, and other unsustainable practices including dumping fish: "[T]rawlers brought in bad fish. . . . One time they used suction . . . to suck it out of the boats. Head gone that way and tail gone that way . . . like soup. Oh it's ridiculous. But the big companies didn't care as long as they got their millions and millions of dollars" (NGP-8, male fisher). Recent improvements, including boxing fish and dockside monitoring, have cut down on some of these quality-related and wastage problems associated with trawler fishing (Fishery Research Group, 1986; Rowe, 1991). Nonetheless, problems persist. Fisher respondents suggested that a sustainable fishery must be one that is based on the production of quality fish, where quality, not quantity, is rewarded financially. Rewarding the production of quality fish means that fishers would be able to harvest less fish to make a decent living: "But I thinks quality is the main thing . . . you only wants half the fish to make the same money" (NGP-8, male fisher).

Elsewhere I show how female plant workers linked the poor quality of incoming fish from trawlers to the production of inferior products, which fetch lower prices than higher-quality products made from fish with superior texture, as well as to wastage of raw material on the production floor (Power, 1997). Producing high-quality fish requires increased time and effort, as well as specific knowledge. Fisher respondents suggested that without financial motivation it is unlikely that this practice would be widely exercised. At the same time, they claimed to be applying the necessary time and effort to ensure the production of quality fish, often resenting the fact that this work was not financially compensated: "I thought I should get more money even if it was a couple of cents or a cent or something. Something for the effort" (NGP-5a, male fisher).

Men fishers often expressed a pride in possessing the skills, knowledge, and abilities to produce high-quality fish. The perception that these skills are being lost with the introduction of new and more complex harvesting and fish-finding technologies makes such claims increasingly significant. According to the "traditional male fisher" model, quality comes from fishers' hands and the tools with which they "choose" to work – this work is embodied in men and embedded in masculine fishing culture. This model is reflected in fishers' sense of an intimate connection with quality, a regard of the treatment of fish on draggers as offensive, and an investment of personal dignity in their product. The alternative organization of work reflected in the "modern male fisher" model and presented by the state and processing companies threatens the current relations fishers have to capital that enable autonomy and the meanings associated with such relations. Pride in work quality and skills, however, was not exclusive to fishers. Men and women fish-plant workers, as well as those men and women who cured salt fish, expressed similar sentiments. Such pride enabled fishers, and plant workers, to make links between ecological devastation and capitalist processes. At the same time, a concern with and the elevation of these skills and abilities had the effect of distancing their work from the practices and intentions of processing companies without effecting change. In fact, it appears that investment in such skills and abilities encouraged fishers to produce better-quality products without financial compensation.

## BLAMING FISHERIES SCIENCE: A FLAWED "LANGUAGE"

Another element of the "modern male fisher" model has been the integration of science at the levels of the individual and the fishing industry. The post-1970s period of the Newfoundland fishery was characterized by the increased presence of a management regime informed by state-funded scientists in the Department of Fisheries and Oceans (DFO). These fisheries scientists relied on an assessment method primarily based on catch and effort data retrieved from the *offshore* commercial fishery and *offshore* research vessels. At the same time, data from the inshore fishery were generally dismissed as being qualitative and anecdotal. Until 1988, this assessment process consistently resulted in excessive total allowable catches (TACs) because of overly optimistic estimates of the biomass of fish stocks retrieved from the offshore data (Finlayson, 1994: 101).

Flawed estimates of biomass reflect problems with the scientific process. First, fisheries science was governed by a techno-utopian approach that assumed sustainable management was possible through the manipulation of discernible variables such as natural mortality and fishing mortality (Finlayson, 1994: 24-5). Second, it failed to account for shifting effort from areas of low to high catch rates, the efficiency of new harvesting technologies, or unreported catches and other unreported activity (Hutchings and Myers, 1995). Third, fisheries scientists tried to balance the variability of research survey vessel data by using commercial trawler data. However, the use of data from commercial fishers is problematic because they do not randomly sample the fish population. Increases in their catch rates can be attributed to greater harvesting efficiency rather than growth in stock abundance (Hutchings and Myers, 1995: 76-7).

Fisheries scientists determined that offshore data from commercial vessels were rational, quantifiable, and comparable to the data obtained from research surveys. Data from the inshore, on the other hand, were dismissed as unreliable, despite the fact that the inshore fishery had contributed between one-third and one-half to the total landings of northern cod (Finlayson, 1994: 102, 105). As such, inshore fishers' knowledge and data were marginalized in the stock assessment process at DFO. Scientists justified these omissions by citing a number of reasons, including "the large number of fishers in the inshore; the complexity of the inshore fishery in terms of gear, local oceanographic variations, and climate; and the absence of any measurement of catch per unit of effort for the inshore" (Department of Fisheries and Oceans, in Neis, 1992: 162). We can probably add a couple of other reasons.

First, the exclusion of inshore fishers and the inclusion of the offshore sector in stock assessment reflect the distribution of and access to power. While the more than 20,000 inshore fishers make up a powerful voting block, the power of the offshore fishery is located in the two main processing companies, FPI and National Sea Products, which wield tremendous political and economic clout (Finlayson, 1994: 104). As Kloppenburg (1991: 524) has argued, accepting local people's ways of knowing could be construed as a challenge to science. Thus, any attempt to integrate inshore data in DFO's assessment processes would undermine the legitimacy of fisheries science. Also, it is likely that the dominant valuations of what counts as knowledge and who are "knowers" mediated the marginalization of inshore fishers and their knowledge. Wright (1995: 132) found negative assessments of fishers' cognitive ability embedded in state documents produced in the post-Confederation period. For example, fisheries planning literature invoked the gender stereotype of the "seafaring man," who was "rugged, independent, fearless, a risk-taker, somewhat anti-social and strong," to explain why a successful fishery would have to be informed and run by experts, not "seafaring" fishers. Fishers' skills and knowledge, acquired through the processes of life and work experience, were not recognized by the state, and thus their warnings of stock declines were ignored.

This exclusion surely helped constitute fishers' understandings of themselves as men and enabled them to construct a sense of themselves in terms of difference, namely between fishers and scientists.[6] Just as fishers used the style of clothes to create a sense of being different from and opposition to the "suits" in political office, they used their "different language" to distinguish themselves from scientists (NGP-7a, male fisher). The common referral to those in positions of power as "suits" and the distinctions made between local and scientific languages highlight power relations and may be read as a form of resistance or opposition to the legitimacy of institutional involvement in the fishery.

The "traditional male fisher" model highlights how fishers attempt to define the self through difference and, perhaps more importantly, the processes through which fishers have been marginalized in stock assessment and management of the Newfoundland fishery. The state's reliance on government-funded normal fisheries science to justify state decisions and policies regarding the regulation of the fishery and the consequent mismanagement have provided the tools that enabled fishers to allocate blame for ecological crisis outside of themselves and to develop a critique of fisheries science. This critique provides valuable insight from a marginalized standpoint and debunks the claim that technology, rationalization, and science are objective applications or pursuits designed to meet the best interests of all members of a given society. Upon examination, it becomes clear that the benefactors are defined along class and gender lines. The present ecological crisis validates inshore fishers' criticisms pertaining to state neglect of local ecological knowledge and the inshore fishery in general.

"THE OLD FISHERMEN KNEW WHAT THEY WERE DOING"

Dominant versions of what counts as knowledge do not go unchallenged. Local fisheries workers tended to invert the official hierarchy of knowledge that placed formal ways of knowing on top. The "traditional male fisher" model challenges state discourses that privilege science, advancing instead fishers' knowledge as the most accurate way of knowing the marine environment. There was strong support among fishers and plant workers, men and women, for the idea that *fishermen* know best in terms of the local marine environment: "I don't think they [the state and its scientists] knows what they're talking about . . . it seems like the old *fishermen* knew what they were doing and knew more about it" (NGP-18, female[7] plant worker).

Exclusion from the stock assessment and management processes, the recent delegitimization of fisheries science with public admissions from the state regarding inaccuracies in the scientific process, and fishers' own observations and experiences have opened up space for fishers (and academics) to construct a critique of normal fisheries science and management. This critique has also opened up discussions on the potential importance of local knowledge for successful resource management and has provided some legitimacy to resource users' knowledge, values, and practices, at least locally and within some academic circles. A recent area of study among academics in Newfoundland is the examination of the content and quality of fishers' ecological knowledge. They argue that because locating fish is necessary to the survival of fishing families, fishers possess extensive experiential and generational knowledge about marine resources and the ocean environment (Neis, 1992; Neis and Felt, 2000; Ommer, 1998). This literature suggests that fishers have developed ways to classify fish and read environmental indicators and have accumulated information about fish behaviour and migration that differs from accepted scientific knowledge and practice.

Using this knowledge, fishers in this study identified some of the shortcomings of science. First, fishers offered a critique of the methods and arbitrariness of fisheries science. Academic critiques of "normal science" argue that because science embraces reductionist and positivist approaches, a holistic understanding of the world is impossible (Harding, 1991; Kuhn, 1974). Instead, "normal science" perpetuates "a hierarchical and linear rather than an interactive and ecological view of nature" (Kloppenburg, 1991: 530). Some fishers echoed this view, though not with such formal wording, advancing a skepticism regarding the usefulness of the arbitrary nature of geographic fisheries management zones that may not reflect the reality of fish migratory patterns:

> Fish got no lines. They don't come down, and all of a sudden, here, coming down around the coast, outside, don't touch the land. This is the scenario that scientists are trying to make us believe. That the fish is coming in from 2J3K, coming down around, without touching land, swinging down around, hits this imaginary line and comes to St. Mary's Bay, and it's 2J3KL fish. It's not real. (TEK-46, male fisher)

Instead of relying on scientific and official knowledge, fishers trust their experiential knowledge to inform them of ecological reality. Indeed, it was through informal testing and daily engagement with their work environment that fishers became convinced that fish stocks were in decline, contrary to the information scientists disseminated. They cited a number of changes that they encountered in their own fishing practices as evidence of stock declines. Inshore fishers experienced declining landings of cod and noticed changes in fish size, despite expansions in fisheries-related work opportunities and increased effort, time, and financial investment by fishers (Neis, 1992: 165). These experiences were inconsistent with optimistic stock recovery predictions by fisheries scientists and pointed to problems with sustainability despite increased regulation of the fisheries and the institutionalization of fisheries science.

Second, not surprisingly given the fact that DFO is state-funded, fishers pointed to the politicized nature of scientific knowledge. Fishers tended to be skeptical of the processes involved in generating fisheries science, which were arguably linked to political and corporate interests: "Over the years, seemed like the scientists were saying what the politicians wanted them to say" (TEK-24, male fisher). Fishers are not alone in pointing to the politicized nature of fisheries science. For example, Finlayson (1994) has documented how social and political processes shaped the production of fisheries "facts" within DFO. Indeed, many of their concerns and connections are paralleled in academic literature that challenges the claim that science is objective and rational (Barnes, 1985; Harding, 1991; Kuhn, 1974; Merchant, 1980; Mulkay, 1979). Interviews also revealed suspicion about the power of individual scientists: "And if they [scientists] made a wrong decision, they were immune to dismissal or anything" (TEK-46, male fisher). Some fishers were concerned about the impact of the pressure scientists experienced to build a reputation on the shape, quality, and content of their research: "I suppose it is like every profession. Some of them [fisheries scientists] is trying to do a real good job. More of them is seeing if they can make a good name or feather themselves. I think the whole society is on the same basis" (TEK-27, male fisher). Some appreciated the contradictory place fisheries scientists hold, pointing to the tensions between doing "good" science and challenging the state and processing companies:

> We get fishermen that's running down science. Usually the people that does it is the loudmouth or the fella because it doesn't work out his own way. I was out on a research boat, just observing, and I think the greater majority of those men, they're out there trying to find the answers. I think some of the things that they are after running up against, if they gets an idea and they wants it to be checked out, now they have got to go against the multinational companies. Therefore, they got to speak against the politicians. So he's there, the politician is there, he only wants to say what he wants people to hear. And let's face it, if you're going to do a good job, you're not going to be very popular. (TEK-27, male fisher)

Many fishers who had involvement in the Inshore Sentinel Fishery expressed positive views towards science and scientists: "I'd say they're doing a number one job with it" (TEK-45, male fisher). The Inshore Sentinel Survey is a fisheries science program, first implemented in 1995, in which DFO scientists work in conjunction with inshore fishers to gather information on the health of groundfish stocks, especially cod (Department of Fisheries and Oceans Canada, 1997). These expressions of approval are positively correlated with personal contact with fisheries scientists through the program, as well as active participation in the union, and negatively with age. Some viewed the Sentinel Fishery program as an indication of the possibility of change in the relationship between fishers and fisheries scientists. This program has provided space for a closer alliance between fishers and scientists working together at the community level since the moratorium, indicating, some suggested, DFO's new-found willingness to involve local fishers in the stock assessment process. Some interpreted the program as an important strategy to gather the knowledge of both fishers and scientists in order to produce a more effective management regime:

> I do think they're [scientists] getting on the right track. Fishery scientists didn't give the fisherperson the credibility of what they knew. We went in to the scientists, and they'd learned the lab procedures and they learned the technology, no doubt about that. But I think they figured, "We knows it all, and you guys, you don't know what you are talking about." I think it's changing. They came now with this Sentinel Fishery. I think this is a step in the right direction. A person, who is involved in a trade, whatever it may be, and he spends a lifetime at it, he might not be able to prove it, or he might not be able to explain it to the sort of people. But the scientists should be open-minded enough to make some sense. With the knowledge they got, and there are things behind it, if they could tie it all together, you could probably make some sense of it. (TEK-27)

Other fisher respondents, though, were less optimistic regarding the impact of the Inshore Sentinel Fishery on the way the fishery is managed. While the intentions of scientists working on the projects might be sincere, some fishers doubted that inshore data would reach management decision-makers or that such data would have any real impact in terms of policy-making even if included in decision-making processes. Some fisher respondents claimed that the implementation of the Sentinel program was a symbolic gesture, designed to give the impression of change without actually altering the existing power and social relations between state management, DFO scientists, and fishers. After all, government officials regulate and control the Sentinel Fishery: "They come out and talk to us but what fishermen is telling them, they don't, it doesn't go back to the minister. . . . Oh yeah, they'll consult with fishermen but when the fishermen says, 'No,' they can't take no for an answer (NGP-2, male fisher).

Others voiced concern that the "co-operative" effort may be used to shift responsibility for management from the government to fishers – a point I will take up in the next chapter. There was also widespread criticism of the accuracy of concurrent test fisheries with respect to the timing and placement of fishing gear and, thus, the reliability of the data collected by DFO: "The surveys, they're going to the wrong areas and the wrong, well the right areas but the wrong time of year. Fish only bes in a certain area a certain time of year and they're gone, right?" (NGP-5a, male fisher). Comments on the unreliability of data with regard to the timing of the surveys were not unusual: "They came out and put crab pots right down next to us and there's no crab there. We knew it was no crab there because at that time of the year the capelin are moving off shore and dying and going on the bottom and they wouldn't pot. . . . So that's when the scientists came out and did their survey, that two days and moved on" (NGP-2, male fisher). Others voiced concerns over the perhaps unnecessary financial cost of these long-overdue surveys:

> Over the last two or three years, they're starting to listen to the fishermen, which they should have been listening to ten or fifteen years ago. Fishermen were crying out then, foul. They should have started then, now they'd have had all this data. Father's generation, grandfather's generation, and his father's before him, they had the pattern of the fish down as good as any scientist would ever get it. . . . So, what more does a scientist want? He doesn't even have to go out there, really. They're spending millions of dollars. I don't disagree with out here, either. There's got to be data and surveys done out there as well as in here. (TEK-31, male fisher)

## "IN THE BLOOD"

Over and over again, the discourses of male fisher respondents explicitly and implicitly linked the fisheries crisis to the ideas, practices, and structures embedded in what I have been calling the "modern male fisher" model. Between the 1970s and early 1990s, this model was used to legitimate a separation between management and resource users, experiential knowledge and science, and sustainable fishing and profit. The problem, by extension, is one of separation and disembeddedness from the "traditional" model. Male fishers described fishing as "in the blood," and thus their desire to "be fishermen" could hardly be regarded as a rational choice. For these men, fishing is more than work; it is a way of life, connecting all aspects of living. Fishing is treated like a vocation. Unlike the shopfloor workers described by Collinson (1992) or the "lads" that Willis (1977) studied, fishers did not separate self from work. Rather, fishing work is deeply connected with who they are, or at least who they wish to be. In fact, fisher respondents found it very difficult to talk about what it means to be a man without talking about what it means to "be a fisherman." Fishers who at some time in their lives spent time away from the sea, either to pursue post-secondary education or through their involvement in non-fisheries paid work, expressed a particular dissatisfaction and returned to fishing: "Well, I went to university first for a year and then I went back to the boats. Fishing, it's in the blood" (NGP-1, male fisher).

This connection to work described by fishers distinguishes their subjective experience of work as primary producers from the working-class experience described by Clatterbaugh, Collinson, Tolsen, Willis, and others: "I threw it [plant work] down and went on fishing. There was something missing and I didn't know what it was until I went fishing" (TEK-18, male fisher). The idea that fishing is in the blood reflects the sense of vocation that fishers feel: "It's in my blood. I've heard of fellas say you get salt water in your blood, and I said, 'Yeah, right,' but it's in my blood. I wouldn't trade it for the world. You're not making nothing there, sure. You're only paying your bills. That's all. Raising your family, but you're not making extra money or nothing" (TEK-33, male fisher). Indeed, some extended this idea to claim that fishing is not only "in the blood," but that Newfoundlanders (again, meaning Newfoundland men) possess a "natural" sense of fishing knowledge because of their collective historical and cultural experiences:

> Nobody has to teach a Newfoundlander how to fish. I always tell everybody that. A Newfoundlander is one person who doesn't have to go to school to learn something. It was something your grandfather did, your father did. It's just a natural instinct. You can pick it up and do it. . . . You pick it up as you go. That's like building a boat. My grandfather built boats, my father never built boat on his own but he worked with grandfather, and I just picked it up and done it on my own, that's all. That's why I always say, if it is something your father did or your grandfather did, a Newfoundlander can pick it up and do it himself . . . it is in the blood. (TEK-22, male fisher)

At the same time, however, the sense of connection to fishing work, described by fishers as "in the blood," presents a unique contradiction. It highlights both the deep satisfaction experienced by fishers and a realization on the part of fishers that, besides fishing work, there is little meaningful work available for men in their communities. In brief, they had no real choice but to become fishers. In addition, when fisher respondents describe fishing as being "in the blood," they reify their positions. In this way, fishing becomes something they were born to do. This sort of determinism, in turn, involves an acceptance of the status quo, including class inequalities and gender divisions of labour, and makes change more and more unlikely. However, it can also be used as a springboard for some sort of resistance to proposed changes to the organization of the fishery.

"THAT'S ALL I KNOWS"

Fishing or fisheries work is hardly a choice in the local economy, where few alternatives are available for men or women – although women's opportunities are particularly constrained: ". . . You know that's here, either you're into fishing on the water or you're into the plants, right? That was the breadwinner here, right?" (NGP-10, male plant worker). For many, leaving their homes and communities is either unrealistic, due to limited opportunities of locating non-fisheries-related work because of a lack of education and formal training, or undesirable, due to a commitment to their families and rural life. One fisher explained that he does not know how to do anything besides fishing: "That's [fishing] all I knows how to do" (TEK-43).

Fisheries workers argued that moving outside their communities to find alternative work proved both impractical and undesirable. One male fisher (TEK-53) decided to return to fishing after a period of land-based work in the nearest city because of a desire to be closer to his family: "What motivated me [to return to fishing] was, at the construction work in St. John's, this being 100 kilometres away, drive to and from every morning. Either that [fish] or spend most of my time away from home, and not see the family grow up. That was a big reason." Locating non-fisheries work outside of the community, then, often translated into long commutes. A major motivating factor in pursuing fisheries work was that this allowed the men to remain in the local rural area: "There were a lot of younger people starting to get involved with the fishery because the way they looked at it, well, the fishery was there and I don't have to leave home and I can make a living" (NGP-2, male fisher).

Material pressures and a commitment to rural Newfoundland kept many fishers in their communities and in the fishery: "He [his father] couldn't work so good as he could, but he managed. I figured I was needed at home more than I was anywhere else. I am kind of a hometown guy. I'm not one for moving around. I'm quite comfortable here. I'd rather be here than anywhere else" (TEK-33, male fisher). Many fisheries workers demonstrated an emotional attachment to maritime culture and valued the benefits of living in small rural communities, such as strong familial ties and the relative lack of crime: "I don't go nowhere. I don't go to St. John's [nearest city]. I only go to St. John's once every three years, but I don't want to be in there no more than that. I like it here . . . I hope I can keep on fishing. I needs it, right?" (NGP-8, male fisher).

In addition, financial investments in homes and fishing property, positively correlated with age, deterred men and women from moving. According to male and female respondents, age acted as a deterrent to returning to school, retraining, or leaving their communities to look for alternative work. The following assessment of moving was typical: "I'm forty, and [my wife] is thirty-eight. . . . Well, it's too late for that now, moving" (NGP-4a, male fisher). Most respondents thought it was easier for younger workers to move and retrain assuming they had relatively little financial and personal investment in their communities and jobs. Interestingly, none of the respondents identified themselves as young workers, even those in their early thirties. From his research on people in the Bonavista-Trinity area, Sinclair (1999; 2002: 304) concludes that age is the variable most strongly associated with migration plans and that "older people, those most satisfied with life in the area, those with children, those with low levels of education, and married people were most likely to stay."

Whether looking inside or outside the local region, most have few other opportunities for employment, having participated in harvesting or processing for most of their lives or lacking formal saleable skills or training in other areas. Fishers reported having little choice about entering fisheries work, often at the expense of their education. Sometimes the question of how and why a man got involved in the fishery seemed like a rather silly question to the respondents. The whole process was rather uneventful, with no real beginning, as demonstrated by the difficulty men had in providing an exact age when they started fishing. Consider this statement by a male fisher (TEK-26): "I can't tell you the exact age, but you know, when you're raised up in a fishing family, you're always in the boat. I was always out in the boat when I was growing up. When I finished school, I'd say seventeen, I went fishing." Another male fisher reported a similar experience: "As long as you're old enough to climb aboard a boat and torment, get in the way" (TEK-27), then you go fishing. Indeed, fishing was often described as less a real choice and more an inevitability.

Personal accounts of work histories, nonetheless, provide a number of insights about how and why men became fishers in a context marked by little choice. First, economic need compelled many households to use child labour – in culturally appropriate, gender-specific ways. Boys, not girls, were recruited to labour on boats. This was the common experience among older respondents, who estimated starting their fishing careers, often without direct payment, as young as age five or six. They were essential participants in the family-based fishery and part of a household strategy whereby survival depended directly on catching and processing fish and subsistence activities. The economic impetus for child labour substantially lessened with the decline of the salt fishery and the introduction of the welfare state, which extended unemployment insurance benefits to fishers and provided transfer payments to households with children, provided the legal requirements regarding school attendance were met (Neis, 1993: 196). This said, younger men also reported beginning their fishing careers as children or teenagers, crewing with their fathers during summer months when school was out. A few men reported being compelled to leave school to fish in the event of death or injury to their fathers, which in turn meant that they had to bear some of the financial responsibility for their families. One male fisher, for example, acquired work on a company-owned trawler when his father died: "I was still in school when father died. Father died, see, when he was only forty-six-year-old. I was only fourteen then. Mother had to take me out of school. . . . There was twelve of us and there was only one working. . . . I had no other choice but to leave. So I been fishing ever since" (NGP-9a). The use of children's labour was not restricted to fishing. It included other fisheries-related work, including making salt fish with other family members.

In keeping with the "traditional male fisher" model, men respondents underscored their abilities to relieve natal households of economic burden by acquiring paid work as soon as possible. For example, one male fisher (NGP-5a) described his teenage years as filled with work – harvesting and salting codfish with his father during the summer months and selling it door to door to acquire pocket money. Because self-sufficiency, materiality, and the reproduction of the household were immediate concerns, educating children was not a priority for most of the families of the men referred to here. This was especially true for older men – one retired male fisher (NGP-6a, age 86) managed to get to grade five before he was needed to fish full-time. The combination of a father's failing health and economic necessity forced many boys to leave school: "Father, before he give up, for four or five years . . . he was sickly, and he wouldn't go along without one of us. I was going to school. 'Don't you sign on no more,' he said. 'You signed on last year, and you signed on again the year. Now, you're not going at that,' he said" (TEK-29, retired male fisher, age 85).

Some evidence suggests that working was regarded as being more important than education in and of itself, outside of material necessity, so that, as soon as he was able, a boy went to work. In addition, fisheries-related work provided an alternative for those boys who did not like or succeed in school: "My son D had grade eleven. . . . He wanted to quit school. . . . I said, 'All right, if you're not going to school, tomorrow morning you gets up and goes in the boat with me.' So that's what he preferred to do. And he's at it ever since" (TEK-43, male fisher). Poverty reinforced an ideology of independence and self-sufficiency, which no doubt mediated boys' decisions to stay in school. This ideology and material realities, combined with a culture that values fishing work, served to reproduce workers for the Newfoundland fishery. Patrilineal inheritance and state policies made the craft easily transferable from father to son.[8]

Willis (1977: 3) has argued that cultural processes equip boys for certain types of work. Certainly the early involvement in fishing work described above helped to shape local boys' understandings of what counts as "men's work," as well as their place in the economy and the value of their work. Like the young men discussed by Willis (1977: 39-40) who took on manual jobs while still in school due to material need, fishers' work as children or as teens – making money in the real world of men – surely provided a sense of confidence, as well as the experience with which they were able to make sense of the world. Although a couple of men respondents did mention active discouragement from parents regarding entry into the fishery because of the hard work it entailed, ecological decline, and poor financial rewards, these sentiments existed alongside of an exaltation of fishing work over other forms of work for men.

### "YOU DON'T LEARN THAT STUFF THROUGH COURSES"

Patrilineal entry into fishing and experiential learning allowed male fishers in this study to flip the dominant hierarchy of knowledge that places formalized ways of knowing the marine environment at the top and local, experiential knowledge at the bottom. Deviance from these aspects of the "traditional male fisher" model was used to explain the unsuccessful management of the fishery. In particular, fishers identified the "miseducation" of scientists and politicians; miseducation was assessed through comparisons with fishers' education. These different educational streams include ideas, beliefs, values, *and* social practices and arrangements, all of which reflect and reinforce class and gender inequalities.

First, many fishers accused scientists of being "too educated" – somewhere along the path of higher learning, scientists lost any intimate connection with the environment or the "real world" that they may have had. According to most fishers, scientists have lost their "common sense" and are out of touch with the workings of the marine environment because of their heavy reliance on formal knowledge, their neglect of knowledge based on experiential learning, and a lack of physical involvement in the actual processes of fishing day to day: "They [scientists] don't have a clue. I don't even know if they know their own jobs" (TEK-15, male crew member on FPI longliner). For fishers, experience enabled the acquisition of skills needed for fishing and knowledge about the local marine environment. Fishers and plant workers, both men and women, generally supported fishers' assessments of experiential and school learning.

Second, fishers tended to construct difference between learning from fishing and the processes of science in terms of ability to account for the unpredictable. Because fishers work in a changing environment over which they exert little control, there is a tendency to embrace and accept the uncontrollable and unpredictable, as opposed to the "techno-utopian" approach of fisheries science (Finlayson, 1994: 24-5). According to Andersen and Wadel (1972b: 153-4), fish harvesting is inherently uncontrollable and unpredictable because a marine environment requires a reliance on limited harvesting and transportation technologies, fish migrate over huge distances, and successful harvesting depends on the particular and changing conditions discernible only when at sea. The belief in the ability to control or account for such unpredictability certainly renders normal fisheries science problematic.

Not only did most fishers reject formal education in terms of relating to the sea, some maintained a superiority over scientists: "The crowd looking after the fishery don't know so much about the fishery as the fishermen know about it themselves" (TEK-29, retired male fisher). Some fishers, differentiating themselves from those who try to understand the sea from books rather than from experience and hard work, assessed their marine knowledge as superior. Consider the declaration – "I'm the scientist" – made by a retired male fisher (NGP-6a). Fisher respondents often insisted that despite their lack of formal qualifications, they would not have made some of the obvious mistakes that the science-informed government has made regarding fisheries management. For example, many voiced concern about an important oversight on the part of fisheries management, namely the impact of a cod moratorium on other species, especially with regard to increased effort. For others, superiority was assessed in terms of toughness to endure the sea. Some accused scientists and others affiliated with the data collection process at DFO of not getting their "hands dirty" (NGP-7a). Still others disparaged DFO observers and monitors, who enforced regulations on fishing boats, for not being able to handle the work. According to one report (TEK-24, male fisher), DFO monitors tended to sleep while placed on boats or worked "between throwing up." Fishers charged fisheries scientists with working in easy and comfortable work conditions, relatively speaking: "The government gave them [DFO scientists] a cruise ship to get aboard, and go out and throw a net out every now and then, and take up a few fish, and come in and say, 'This is what the stock should be here'" (TEK-46, fisher, age 38).

Contemporary younger fishers were more likely than older respondents to have participated in some type of formal fisheries-related or water-safety courses both before and since the moratorium. This likely reflects an original, but later withdrawn, criterion of TAGS that defined fishers as active only if they participated in professionalization efforts, especially training (Human Resources Development Canada, 1998: 13), and thus might indicate another disjuncture between fishers' beliefs and behaviour. It might also signal a shift in the hegemony of models at the local level away from the "traditional male fisher" model and towards the "modern male fisher" model – a shift I explore in greater detail in the next chapter. Nevertheless, while most fishers conceded that education and training courses may be appropriate for trades, there was a strong tendency to dismiss the usefulness and practicality of such courses and training in the day-to-day catching of fish:

> To be honest with you, you don't learn that stuff through courses. That got to be bred into you, more or less. It's not like going to learn mechanic's job. Every day there is something different. I can tell you how to set a particular trap but that's no good. You have to be there in that particular place, you got to know what the bottom is like, what the tide is. It's a job to explain it, you've got to have a feel for it. (TEK-44, male fisher)

This comment supports the idea that the acquisition of practical knowledge and fishing skills is a "bodily exercise" (Palsson, 2000: 37). Such acquisition involves an integration of the generational transfer of knowledge, the immediate task, the natural world, and social relations. Hands-on experience tended to be more highly valued than formal qualifications: "Yes, everything I did, it was hands-on from him [his father]. That's the only way, I think, not the only way, but the best way of learning, is hands-on. And I also went to school after and took some training in navigation and engine repair and stuff like this" (TEK-26, male fisher). Younger men regularly praised older fishers, who are "uneducated" but smart in the ways of the sea, for being the true scientists of the water. Expressions of respect for older, "uneducated" fishers were common:

> Well, a lot of fishermen around got no education but they understand what's going on. Probably if they had a chance to go to school now they would be scholars. There are a lot of them around like that. There is a fellow over here . . . just got out of school and went on doing his own thing, and done the best kind as a fisherman. His young fellow now, he's doing best kind because his father taught him good. (TEK-44, male fisher)

Paradoxically, formal sea-related training appeared to enhance the status of some fishers. For example, one male fisher (TEK-27), who earned a Merchant Marine certificate and a Nautical Science diploma from the Fisheries College, obtained his position on the vessel because of formal qualifications: "I was the skipper because the rest of the family, they didn't have any education. They had local knowledge though." Other fishers suggested that book learning complemented experiential learning: "I picked a lot of it up on me own and I read a lot of books and that's basically how I got it. . . . And trial and error but that trial and error racket is costly" (NGP-5a, male fisher). The difference here, though, is that this fisher taught himself through books rather than being taught in a formal institution. Nevertheless, it was in these discussions about training and education that an acceptance, even if reluctant and partial, of the "modern male fisher" model emerged – a model that links fisher status to among other things, formal qualifications.

Another way in which fishers differentiated their "education by fishing" from the formal education of scientists pertained to the customary generational transfer of oral ecological knowledge from fisher fathers to sons, including family fishing secrets, information about fishing grounds and berths, various techniques for catching fish, and such navigation skills as using local landmarks. Despite an increase in the practice of recruiting non-kin as crew members since Confederation, patrilineal recruitment and inheritance of fishing knowledge and property characterized the experience of most of the male fisher respondents described here. As young new entrants, boys entered into apprenticeship-like relationships with older men, usually fathers, who passed on valuable fisheries knowledge and skills in male spaces, including wharves and boats (Silk, 1995: 265-6). This method of learning about fishing was widely reported: "I used to fish with my father so much, just go out and sound, sound fishing and gill nets, when I was younger, right? All when I grew up . . . off and on, right? Grew up alongside of the water and we likes the water" (NGP-8, male fisher).

Fishers reported that there was no or little systematic teaching by fathers. Rather, learning occurred by observing the goings-on in the world of fishermen: "There wasn't any actual teaching I guess, but just what I picked up from Father. . . . Even before I got out of school I always fished. When I was even younger, summertime, that's what I was always at. I was always around the water. Ever since I can remember" (TEK-33, male fisher). The non-direct teaching and learning relationship between men and boys appears to have been commonplace: "[A]s far as I'm concerned, when you grows up and your father's a fishermen and you're down around the wharves and down around the stages and everything, so I thinks myself that it just grows on you, right?" (NGP-9a, fisher, age 44). No doubt, along with learning practical ecological information and fishing skills, boys in this relationship are subjectively prepared for a particular type of work. Learning to fish is accompanied by learning that fishing work demands a particular approach to life, as well as being a particular type of man. In this way, fishers' investment in informal learning mechanisms and the praxis of fishing are about more than practical solutions to the immediate problem of making a living. Rather than separating work from the self, which is a common working-class theme (Clatterbaugh, 1997; Edley and Wetherell, 1995; Willis, 1977), fishers understand fishing and how they learn to fish as being intimately connected to the type of men they are.

Closely connected to these experiential and generational learning processes is the sense of vocation that fishers feel. Connell (1995: 170) has argued that expertise can be defined in different ways. One way emphasizes formal qualifications to demonstrate intellectual ability. The other underscores vocation, or to put it another way, the sense that "[t]he whole person is engaged in the work." The latter approach best describes fishers' definitions of expertise. The vocational commitment to fishing, combined with a higher valuation of practice and patrilineal transmission of knowledge over school-based learning, provided fishers with a sense of being able to get at the "truth" about fish behaviour and the marine environment in ways that science cannot. This subjective claim to be able to get at the truth parallels a working-class theme noticed by both Willis (1977) and Collinson (1992). Moreover, this claim reflects a particular reality. Indeed, formal fisheries science has been undermined to some extent with ecological collapses. However, Collinson (1992: 78) reminds us that the elevation of a particular type of male work is about more than getting at the truth. It is also about enhancing personal status and obtaining a positive, masculine self-image. Understood this way, fishers' attempts to invert the dominant hierarchy of knowledge and to use discourses that underscore the importance of experience are strategies that compensate for and reflect their lack of control in the direction of the fishery, as well as a presentation of an alternative way of determining worth. This alternative, as has been emphasized, favours experience and being satisfied rather than education and the pursuit of profit. Fishing is subjectively experienced as "truth," which discredits official versions of the reality of the health of fish stocks and dominant evaluations of work, education, and status. In this way, local discourse is an important sign of social identity.

The alternative, and oppositional, approach to assessing worth celebrates experiential and patrilineal knowledge and reflects fishers' contradictory class position. Yet, this approach does not necessarily translate into a movement for change. Respondents overwhelmingly felt that had the state listened to fishers' warnings regarding the health of fish stocks years ago, things would be different today. Because of their experience and intimate ecological knowledge, fishers insisted they possess insight into the requisites for a sustainable fishery and, thus, should have some say in management: "We should have one hundred percent say in it [management]. We wouldn't be in this mess now . . . because they [fishers] fish, and they know what's going on. They know the waters. They know what's there" (TEK-13, male fisher).

However, fishers' insistence on their inclusion in management is not necessarily a radical perspective. Fishers often resigned themselves to the understanding that "politics" will always prevail and determine the fate of the fisheries. While many fishers did challenge fisheries science and the reliability and impact of such co-operative efforts as the Inshore Sentinel Fisheries, they did not tend to radically question the existing relationship between fishers, management, and science. Fishers insisted that they should be listened to by DFO, but failed to challenge state control more fundamentally by imagining a fisher-run industry:

> [F]ishermen should have an input, right? I mean you just can't leave the scientists out of the picture either. Their technology and how they does their assessments is definitely needed, you know? DFO still got to have their power because, I mean, if you leaves it all to the fishermen, the fishermen in one sense is going to strain it all too because they're out making a dollar. I think if you leaves it all to the fishermen, the fishermen should have their input, definitely, but I mean that's only to a certain extent I would say. (NGP-1, male fisher)

This position may reflect a partial acknowledgement of the usefulness of science and certainly reflects an acceptance of the tragedy of the commons thesis. However, fear of radical reform may also be explained in terms of allocating blame. A fisher-run, community-based management strategy that lacked the proper resources would inevitably fail, transferring blame from scientists and government to fishers. This position appears warranted given recent state-directed privatization efforts that download more of the costs and responsibility to fishers. Closely connected to this is fishers' recognition of the part they have played, however small, in ecological declines. Some fishers conceded that they were also responsible for mismanagement and scientific uncertainty by providing inaccurate information to scientists and the government in order to protect their own interests or by hiding illegal activity, including under-reporting. These practices have been documented by academics (see Hutchings and Myers, 1995) and recognized by respondents:

> I think a lot of problems with it is the accuracy of the methods they [scientists] was using to get their information. There again, another problem was with the misreporting that the scientists were getting, from fishermen and companies. Anybody can be as good as they like, but if they are given wrong information, there is no way you can make a sound judgement on that. (TEK-22, male fisher)

However, as mentioned earlier, fishers rationalized such deception as an economic strategy. Playing by the rules, created by the state, did not pay off.

The difficulty in imagining a fisher-run management regime also reflects a general unwillingness to share local ecological knowledge. This is not surprising given its important role in determining ability to catch fish, which is extremely important in times of ecological and economic uncertainty. Safe and inclusive spaces for open discussion might correct for such examples of misinformation, but this will be a difficult process given the history of distrust and marginalization. Withholding information from DFO and not co-operating are also indicative of efforts to resist and take back or retain some control, albeit on an individual rather than collective basis. Of course, the contradictory position put forward by fishers – claiming to possess knowledge about the reality of fish stocks and the marine environment yet rejecting the idea of a fisher-run management regime – is supported by a celebration of fishing as being "in the blood." Because fishing is highly valued and deeply connected to their understandings of themselves as men, becoming involved in the management within the existing regime, is unthinkable. Yet this rationale reproduces the status quo and fails to question the basis of their relative powerlessness.

## SUMMARY

A widespread theme in the interviews with male fishers was the allocation of blame for the latest fisheries crisis. Using the "looking to the past" framework and the "traditional male fisher" model, fishers were able quickly to identify the state, its science, and business as the primary culprits. Some fishers blamed all three; others identified one or two as the key wrongdoers. Simply stated, the cumulative argument put forward reduces the decline of fish stocks to the pursuit of profit, the skewed power relations between the actors in the industry – favouring big business – and the lack of experiential, hands-on learning on the part of those doing the science and making the decisions.

When people are able to connect their problems to society and have access to alternative discourses – like the inshore fishers here – they are more likely to engage in political social action. Local discourses that privilege fishers' knowledge enable the construction of oppositional explanations and culturally meaningful male identities and constrain participation in formal training. Despite the real insights fishers offered regarding the health of fish stocks and the relations among the major actors, spaces for individualism and divisiveness likely serve as impediments to collective action. The main reason for this is that the traditional model in which their responses are based is simultaneously subversive and reactionary, egalitarian and individualistic.

The "traditional male fisher" model helps point out who and what are to blame, but it stops short of developing an alternative to the current social, political, and economic arrangements. This model challenges the hegemony of the modern as a superior way of knowing, ruling, producing, distributing, and consuming – as exemplified in the state, science, and capitalist arrangements; yet, at the same time, the model reproduces the structural position of fishers within those arrangements. Fishers often disagreed about the value of science and the extent to which inshore fishers themselves are to blame. Some of these debates are dedicated to portraying oneself in the most favourable light possible. When fishers accuse *other* fishers of being partially responsible for the fisheries crisis, they tend to do so by using themselves as the standard by which others should be measured. Similarly, debates over the usefulness of science in fisheries management may reflect a desire to present oneself in a particular way, as perhaps more "enlightened," more "modern." In this example, the "traditional male fisher" model that emphasizes experiential learning is challenged, even if partially. After all, many fishers engaged in such debates simultaneously held the ideas that fishers know best, that some fishers acted irresponsibly, and that science is valuable.

These individualistic and divisive responses reflect fishers' concerns about their future involvement in and confusion over the state's direction of the fishery and fishers' unlikely possibilities of locating alternative work in urban areas, given their low educational levels or particular skill sets and their commitment to rurality. The responses indicate fishers' awareness of a shift in values informing state direction – from an emphasis on science to one based on classical economics – and fishers' agency in their decisions to accommodate, or not, such a shift. Such decisions are shaped by fishers' present and perceived future material conditions and are enabled by their position in production. These decisions help to sustain and transform cultural values, norms, and practices.

# *Struggling with Enclosure* 5

## "ABOUT TIME THEY RECOGNIZED US"

The scientific discourse that has guided fisheries management in Canada is being replaced by a neo-liberal discourse that underscores free-market capitalism. Like the scientific discourse, however, it derives power from its exclusionary processes that define experts and non-experts. In the paradigm that currently dominates the state's plans for the restructured fishery and its participants, the experts are economists instead of scientists. In this version of the "modern male fisher" model, corporate interests take centre stage. The objectives are to reduce capacity, rationalize access, and create self-reliant and self-managing individual entrepreneurs (Department of Fisheries and Oceans, 1996, 2001).

> Government will need to continue to be involved in determining what constitutes "best use," but over time, DFO wants to remove itself from decision making concerning commercial allocations of the resource. A condition for DFO removing itself from allocation arrangements will be the establishment of appropriate rules and documentation of the shares held by individuals and fleets . . . . (Department of Fisheries and Oceans, 2001: 34)

In an effort to transfer more responsibility to participants, the federal government has instituted enterprise allocations, individual quotas (IQs), and individual transferable quotas (ITQs) that allocate quotas to individual enterprises or harvesters, thereby replacing annual renewal arrangements with longer terms for fleet shares and allocations (ibid., 13, 31, 34). With enterprise allocations, processing companies with offshore fleets are assigned part of the annual quota. Individual quotas in Newfoundland take the form of boat quotas, in which individual boats are assigned part of the quota and the rights are not transferable or divisible (McCay, 1999: 301). The federal government's restructuring strategy also involves a reclassification of inshore fish harvesters into core and non-core categories. Exclusive membership in the core group, entry into which is through replacement only, aims to further limit capacity. Prerequisites for membership into the core group include heading an enterprise, holding key licences, and demonstrating attachment to and dependency on the fishery (DFO, 1996; 2001: 24). The core and non-core categories are used by the state to determine who gets what fisheries resources. For example, core members have privileged access to new licences, and replacement vessels.

The implementation of core and non-core statuses and efforts to shift governance have occurred alongside provincial and industry initiatives to professionalize fish harvesters (DFO, 2001: 98). Newfoundland and Labrador was the first province to legally recognize and promote fishing as a "profession." It has done so through the establishment of the Professional Fish Harvesters Certification Board and the Professional Fish Harvesters Act. Together, the Act and the Board entrench a set of criteria – apprenticeship programs, formal training, and experience requirements designed to limit access and keep out "moonlighters" – to improve the status of the occupation and to increase a sense of pride, security, and opportunity among the remaining participants. Opportunity and security would be commensurate with a fisher's advancement along the Professional Fish Harvesters Certification Board's gradient designations – Apprentice Fish Harvester, Level I, and Level II. These designations, in turn, reflect the completion of government-recognized training and experience. In 1997 the registration system used by DFO incorporated and applied these designations, with existing fishers grandfathered to the appropriate certification levels based on historic attachment, defined as the number of years fishing, and dependence on the industry, defined as fishing income. These policies fit with the broader state direction that has offered restricted access, exclusivity, and privatization as solutions to the current environmental crisis.

Other organizations aimed at realizing professional status for fishers take a critical view of the current direction being followed by the state. For example, in its response to the federal government's Atlantic Fisheries Policy Review, the Canadian Council of Professional Fish Harvesters (CCPFH, 2001) challenges DFO's solutions to stock declines and overcapacity through privatization and offloading of responsibility of allocation and conservation on the industry. In its report, the CCPFH maintains that the federal minister must have ultimate control to protect accountability, which is lost through privatization. According to the CCPFH, sustainability can be achieved by linking rights of access to owner-operator fleets and limiting the movement of corporate capital into the inshore. The failure of the Atlantic Fisheries Policy Review to address the legal loopholes that have eroded the owner-operator and fleet separation policies is cited as a cause for concern. These loopholes allow companies and non-harvesters to buy up licences and quotas legally through "replacement licences" – a process that has been upheld in courts. Although transferring a licence is illegal, it is common practice through the "replacement licence," which has been used to uphold contracts between fishers and companies and allows the use and title of the licence to be separated, essentially transferring the right to use the licence to the processor while keeping the title in the fisher's name. The company can then insist that the fisher request a replacement licence for an eligible person designated by the processor (CCPFH, 2001).

Concerns about the allocation of and access to fisheries resources are justified. Research on fisheries in Canada and elsewhere (Neis and Williams, 1997; Skaptadottir and Proppe, 2000) indicates that once quotas are transferable, individuals and corporations buy them up, leading to a concentration of ownership and the elimination of rights of crew members and their families to independent access to such resources. In addition, as Neis and Williams (1997: 54) have pointed out, "ownership" of resources does not necessarily mean conservation. While IQs will limit competition temporarily, small-boat quotas, high costs to vessel owners for fisheries management, higher licence fees, and more limited access to employment insurance (EI) benefits will push fishers to increase effort in the long term. Transferable IQs have been associated with ecologically adverse effects (Copes, 1996). Participation in the fishery will increasingly be determined by economic means. The push to reduce industry capacity by limiting membership is based on the assumption that there are too many fishers and not enough fish. This assumption ignores the differential impact of the various technological and harvesting capacities between the inshore and offshore sectors, and it ignores the fact that the most serious stock collapses have occurred where there has been heavy regulation and where there has not been open access to the commons (Alcock, 2000).

Nevertheless, there is broad acceptance in Canada of the need for fiscal restraint, especially when dealing with the Newfoundland fishery. This largely reflects the view that fisheries workers are a tremendous drain on the social welfare system (Neis and Williams,

1997: 48). Acceptance of this view has allowed the federal and provincial levels of government to go ahead with a neo-liberal agenda that includes downsizing the fishery – and the infrastructure in place to manage it – and establishing new exclusionary criteria for access to resources. Inshore male fishers have been engaged in these processes with varying degrees of accommodation and resistance. The continued scarcity of groundfish, combined with the overfishing of other species and the cuts to adjustment benefits, contributed to growing support among fishers for a more limited, professionalized fishery. At the same time, they have resisted downsizing. This chapter examines the impact of and reaction to the current movement towards enclosing the commons. The current criteria used to determine legitimate harvesters and the methods of distributing access to resources via individual quotas have served to reinforce a history of male privilege and to solidify corporate interests. The state's direction both deviates from the earlier scientific-based federal management model and has been enabled by it.

While fishers often articulated clear judgements about who and what were responsible for overfishing and stock declines, they tended to be less certain about exactly what should happen next – as indicated in their ambivalence regarding the future relationship between fishers and scientists. There was a fair degree of support for professionalization efforts pursued by the union and endorsed by the provincial and federal governments. Some were very enthusiastic: "About time they recognized us" (NGP-5a, male fisher). Fishers' support for professionalization can be explained at least in part by the shared sense of stigma among Newfoundland fisheries workers. Newfoundlanders generally are proud of whatever it is that makes them distinct from other Canadians. At the same time, they recognize that this differentness is used to degrade and stigmatize Newfoundlanders. Further still, fisheries workers see themselves as being distinct and different from "townies"[1]: "I think that a lot of people . . . looked down on fishermen" (NGP-5a, male fisher). Those who worked in fish-processing plants also felt such stigma and defended this work against stigma in terms of money: "We felt a little bit inferior because we worked in a bloody fish plant . . . I'm sure there's people gone into other jobs now and never made half the money that you made up there" (NGP-13a, retired plant manager).

This sense of stigma was intensified with the implementation of financial compensation packages and other adjustment programs that effectively invalidated their work. A decline in the standard of living, reliance on government support, and popular misunderstandings regarding financial compensation packages and retraining opportunities added to this stigma. Men and women, fishers and plant workers, recalled stories and personal experiences demonstrating the stigma attached to NCARP and TAGS and suggested that "townies" and mainlanders "act superior" to fisheries people (NGP-7a, male fisher). To paraphrase one male fisher (NGP-7a), outsiders perceive fisheries people on TAGS as undeserving, lazy, and getting a "free ride." Plant workers were not exempt from the degradation attached to being a TAGS recipient:

> People believes you're on TAGS you got it made. I know a lot of people do. . . . But how can you be, when your salary is cut in half, and your kids are looking for this and that and you can only give 'em probably the bare necessities, food and dressing now, you know what I mean? Like right now I needs a new car and I can't do it. . . . Everyone needs it (TAGS) but nobody wants it if there's an alternative. (NGP-13a, retired male plant manager)

Most fisheries workers sensed that outsiders lacked an understanding of their reality. But the lack of understanding did not stop there. A female plant worker (NGP-12b) recalled a story about a man's daughter who was ashamed of him because he received TAGS. Negative images of fisheries workers are perpetuated in the popular media (Robbins, 1997) and local fisheries workers were well aware of this: ". . . turn on the open line shows and you can hear it. Pick up the newspaper. I mean all they [non-fisheries people] say is, 'What do these people want? What more do they want?' We don't want this. We didn't ask for this, right?" (NGP-17, female plant worker). Some respondents understood the hostility expressed by outsiders towards fisheries workers: "People resent us, I believe, for getting [TAGS], I suppose for so long as we're getting it. I agree with them to a certain place, you know? It's not going to last forever" (NGP-14a, male plant worker).

Professionalization, then, was interpreted by some as a challenge to such stigma. Yet professionalization is available only to fish harvesters, not processors; and it poses particular problems for female fishers (discussed in the next chapter) and offers a direct challenge to many of the aspects of the "traditional male fisher" model and the "looking to the past" perspective that were concurrently valued by most fishers. Fishers tended to reject the idea that fishing could be learned through formal curriculum or schooling and to define expertise as being grounded in experience and acquired skills. They were skeptical about academic learning.[2] Fishers laid claim to "common sense" based on what they learned from their day-to-day work experience, which had correctly discredited formal versions of the health of fish stocks before the moratorium. A part of learning to fish involves a tactile, hands-on dimension, which makes knowledge difficult to quantify or easily understood for someone without this first-hand experience. This qualitative aspect reflects the observational approach to learning, as opposed to following explicit verbal or written instructions. Others expressed ambiguous feelings regarding the meaning of professionalization and reduced it to a change in talk only: "It's all right I guess. But like, I don't [think] there's any changes" (NGP-4a, male fisher). To some, professionalization appeared to be an inevitable part of "progress": "I suppose it's okay, you know? It's only trying to improve things, you know? You got to change with the times" (NGP-8, male fisher).

In discussions of professionalization and the closing of the commons, the alternative and competing approaches to "looking to the past" and "traditional male fisher," i.e., the "modern male fisher" model and the "forward-looking" perspective – first encountered in discussions about assigning blame – were more fully developed. Fishers adopting this latter perspective were preoccupied with establishing their occupation as a "profession" and limiting access. In this chapter I examine how these discourses are used by fishers and the state to create the new requirements that define who fishes and under what conditions. While some fishers unequivocally aligned themselves with one model or the other, many male fishers used both arguments to make a case for their inclusion. This reflects perhaps a lack of clarity in the meanings of core status, professionalization, and the economic discourse of privatization, as well as fishers' contradictory class positions.

## "WHAT DO THEY CALL A FISHERMAN?"

The context of uncertainty and scarce resources and the state's exclusionary rules, which appeared ever-changing to fishers – a problem recognized by DFO (2001) – have fostered individualism and divisiveness among fisheries workers and in communities. Based on their observations and experiences, many male fishers decided that DFO rules and policies, such as the new core classification schemes, arbitrarily and unpredictably excluded "genuine" fishers from practising their craft and making a living and cut fishers prematurely from financial compensation (Human Resources Development Canada, 1998: 38-41). Fishers complained about not knowing or being confused by the "rules." Would they be next to lose access to government assistance or to be refused a licence?

According to fisher respondents, official definitions of fishing status used in the core classification scheme did not make sense to them. These definitions ignored the social and material reality of their lives and disregarded relevant information and the informal social relations of crews. These omissions had negative implications for small-boat fishers in terms of their ability to acquire licences and financial compensation: "I was exempt from any schooling, exempt from anything, because I was fishing so long. Then when they [the government] changed it over to TAGS, I don't even meet the criteria. . . . Which fishermen are they picking out? What do they call a fisherman?" (TEK-19, male fisher). This experience of uncertainty and the recognition of the very real power of the state to determine the direction of and membership in the future fishery have created much anxiety among small-boat fishers about their individual futures and have limited any sense of collectivity beyond membership in a particular harvesting sector.

At the time of the interviews, few accepted the idea that total privatization is the best way to ensure conservation. However, as recognized by Neis and Williams (1997: 54), declining stocks and incomes make individual quotas and perhaps individual transferable quotas appealing to fishers – the former secure access to fisheries wealth; the latter enable fishers to sell and exit a declining industry. There was widespread support for the state's position that there are "too many fishers and not enough fish." The problem with this assessment, as discussed earlier, is that it ignores the technological differences between a small-boat fishery and a company-controlled offshore fishery. Some fishers, assuming that greed is part of human nature, accepted the tragedy of the commons thesis and argued for limited entry. In addition, most fishers believed that boat quotas encourage safe fishing practices. According to fisher respondents, boat quotas eliminate risk involved in a free-for-all and provide fishers with a better chance of making a decent living. It was argued that boat quotas allow fishers to remedy gear or technological problems without losing valuable time to catch the allotted quota and to fish more safely:

> There's no big rush, you haven't got to go out in all kinds of weather. We was out there twenty miles last year. Sure it wasn't fit. There was boats out there, fifty-five-footers with stabilizers on them, and they had it rough enough. And we only had a thirty-five-foot boat. We had to go, we never had no other choice. If we didn't go, we would have lost $9,000 or $10,000. . . . The boat quota, you can get the weather, take your time and go and get it. It's too big of a rush. Buddy lost overboard there last year up there, he had to go. Another fellow, he broke his neck on the boat there last year. He was out, storm wind, he fell out of a chair or something, went down and broke his neck. (TEK-28, male fisher)

Support for some state enclosure policies may be explained in terms of fishers' concern with egalitarianism. On the one hand, most fishers agreed with the assessment that boat quotas are a fairer management strategy than the allocation of overall quotas. Boat quotas ensure that all fishers get a piece of the pie, even if the pool of fishers is shrinking. In her study on the northeast Newfoundland crab fishery, McCay (1999: 311-14) argues that local people's resistance to ITQs and their acceptance of boat quotas reflect cultural concerns with distributional equity and morality. Locals believed that scarce resources should be spread as widely as possible. ITQs, on the other hand, are attached to private and tradable property rights and they risk shifting control and ownership from local communities to outside and transnational bodies (McCay, 2000).

Others believed that the supposed problem of "too many fishers and too few fish" necessitated state management. In this case, a fatalistic approach, which holds that the direction of the fishery is out of their hands, and a cynical view of human nature, which supports the assumption that left to their own devices fishers will destroy marine resources, have led fishers to endorse the belief that outside protection of the commons is necessary: "[I]f you got to leave it up to the fisherman, I don't think it will be a very good fishery" (TEK-41, male fisher). Another male fisher stated: "Especially when you got numbers . . . everyone trying to get their hands on certain quotas and what not and the fish is not there. I mean it's not realistic. But if DFO regulate it to a sense where they give boat quotas or whatever for the amount of fish that's there, well hopefully you be able to make a living" (NGP-1, male fisher).

Both "solutions" – neither of which question the legitimacy of the supposed numbers problem – encouraged fishers to engage in individualistic attempts to secure a place in the fishery in their assessments of who is and is not a "genuine fisher." In response to government policies and regulations that focus on defining who should be allowed to continue to fish and who gets access to quotas, fishers differentiated themselves from others. In response to economic and symbolic vulnerability, fishers attempted to prove their authenticity by elevating themselves over others. To prove why they should not be removed or displaced, men fishers created dichotomies between genuine or legitimate fishers and non-genuine or illegitimate fishers. Fishers tended to quickly draw the line at their feet regarding legitimacy and developed a set of criteria for inclusion around their personal work histories and criteria for exclusion around others by using – often simultaneously and contradictorily –

the "looking to the past" and "traditional male fisher" frameworks and the "forward-looking" and the "modern male fisher" discourses.

Using the "looking to the past" discourse, many fishers characterized "real fishermen" as those with long careers on the water, accumulated experience and skill, and possession of traditional ecological knowledge passed on by fathers – criteria that pose obvious problems for female fishers as new entrants on male partners' boats. Emphasizing attachment to traditional methods gave these fishers the ability to lay claim to "traditional rights" – meaning adjacency and historical dependency criteria, some versions of which have been used by the state to determine access to fisheries resources and adjustment programs and financial packages. As such, fishers espoused this perspective in developing arguments that supported their involvement in the future fishery. That said, most recent state documents have revisited the traditional rights criterion, and it is clearly being reworked to fit the male business model that emphasizes rationality and capitalist accumulation.

Some fishers used these claims and comparisons with other fishers to secure dignity and critique their exclusion from financial compensation or the procurement of access to resources. Consider how one small-boat male fisher (TEK-32) objected to the current crab quota allocation process for failing to consider "traditional fishing attachment," which he claimed led to his exclusion: "I can't even fish a crab and I have been at this for twenty-two years." As far as men fishers were concerned, "bona fide" fishers should have priority to certain waters and species because of their traditional involvement. Respondents pointed to many examples of "legitimate" fishers falling "through the cracks" and, as a result, being cut from financial compensation programs or being denied a particular fishing status needed to acquire certain licences:

> There's some guys who applied for this [core status] and been fishing for years, probably seventeen, eighteen years, and never got it because of technicalities. It's based on years fished . . . but you got some guys who had a groundfish licence and fished on longliners with other enterprises, and those guys . . . fell through the cracks. Some guys, like there was two in a boat and, like myself, all the fish was put in my name, the receipt in my name. And me brother, who fished with me, had a groundfish licence but never used his own small boat, fished with me, and those guys fell through the cracks. (NGP-1, male fisher)

Another way in which fishers justified their inclusion in a downsized fishery was by using a "forward-looking" perspective. Fishers taking this approach tended to be younger and demonstrated commitment to the fishery by emphasizing their willingness to "change with the times" – measured usually in terms of accumulation of fishing gear and communication and labour-saving technologies, movement into larger boats, and overall increased financial investment. For example, when asked about changes in fishing technology over the years, one male fisher (NGP-9a) was quick to point out that he had invested in technology sooner rather than later: "No, but it didn't take me long to get it though." Fishers exercising this perspective often adopted the lingo of the assumed rational man in neoclassical economics, using words like "enterprise," "vessel," and "harvester" instead of "livelihood," "boat," and "fisherman" to describe their position in the industry, which had the effect, even if unintentional, of denying the cultural context and complex social relations in which their work takes place. A common interpretation of state fisheries policy among locals was that larger boats (over thirty-five feet) are favoured in the quota and licence allocation processes and thus there is tremendous pressure to choose between "going bigger" or getting out. Fishers who took this former direction tended to describe it as a strategy to deal with stock declines, which necessitated increased effort. This act and the willingness to adopt new fishing technologies are also intended to symbolize a present and future commitment to the occupation.

The shift in emphasis from belonging to a fishery conforming to the "looking to the past" perspective to belonging to a modern, technology-driven industry may reflect pressure to conform to dominant notions of what it means to be a "modern male fisher." The concern with appearing compliant and compatible with the dominant versions of who and what should and will be included in the fishery of the future, of course, reflects a recognition that government has the power to exclude and include. For some, moving into larger boats and procuring new equipment and technologies were acceptable trade-offs; sacrificing some of the culturally valued aspects of the "traditional" fishery was intended to preserve their participation in an insecure fishing industry. This move did not necessarily reflect, then, fishers' beliefs about how the fishery should be run. Regardless of ideological commitment to the "forward-looking" perspective, it takes time and money to procure technology and there are restrictive requirements, leaving some fishers behind. Newer entrants had particular difficulties financing new technologies and gears for their boats, leaving them in risky positions: "Takes a few years to get into it" (NGP-8, male fisher).

Investment in and the adoption of technologies are not new.[3] Fishers' work histories report lengthy lists of abandoned and replaced technologies. In fact, most of the men lauded artifacts with labour- and time-saving and safety-enhancing capacities. The adoption of technology eased manual work: "Now you got the hydraulic stuff. You haven't got to work" (TEK-48, male fisher). Technology is used to replace crew: "It is better than a dozen men because they get tired after awhile. And you've got to feed them and listen to them grumbling" (TEK-44, male fisher). According to fisher respondents, the adoption of multiple technologies like fish finders to track and locate fish, radios to aid in navigation and communication, radar and loran, and the replacement of wooden boats with fibreglass boats instill confidence and reduce the perception of risk because "it is not all going to give out the one time" (TEK-18, male fisher).

Others stressed how new technologies made harvesting more efficient: "The Loran C, that was the cat's arse, we say. Once you got to know how to plug in the numbers and understand the gear, no problem at all. You could go anywhere by the ocean. You could go where you liked then. That's when we started moving" (TEK-27, male fisher). Some younger fishers suggested that new technologies such as global positioning systems, loran, and radar took the guesswork out of traditional inshore methods, including relying on landmarks or "rocks" to navigate: "Father used to tell me the landmarks. I couldn't remember them all. . . . Well, rocks is not all that accurate. . . . You have bad days and then you wouldn't be able to see nothing . . . and plus getting home out of it on foggy days. One time you use watch and compass . . . and that was kind of hard. You didn't feel comfortable" (NGP-5a, fisher, age 33). The acquisition of fish-finding and navigation aids provided an increasingly needed competitive edge: "Just about every one of them [boats] got a Loran C or GPS [global positioning system] aboard. They miss very little fishing time now, unless there is a gale of wind or something. Fog and stuff don't bother them. I could be out there half an hour looking for a buoy, but they are right on their gear. That's where they gain the time on a fellow like me who's got neither one of them" (TEK-31, male fisher).

While they recognized advantages to the adoption of technologies, inshore fishers, young and old, were critical of the impact of these technologies on the declining fishery resource and traditional skills related to the inshore fishery. Fishers voiced these views whether or not they employed such technologies: "Technology has got to the point that the fish can't even find a place to hide away, now. Technology has got to the point, if there's one fish there on that fishing ground, and if I steam over, anywhere in a certain radius, I'm going to find that fish. That fish don't stand a chance" (TEK-31, male fisher, age 40). The reduction of the manual labour and risks entailed in fishing was met with ambivalence, especially by older men fishers and those invested in the traditional male fisher model. This ambivalence reflects, at least in part, the concern that the perceived improvement of work conditions might somehow lessen or redefine the "masculine" nature of fishing work. Older fishers' investment in the physical aspects of fishing was mediated by their work experiences in the pre- and immediate post-Confederation period of salt-fish production. During this period, fishery-dependent families relied on household members for survival, not the welfare state, and fishers relied on their own physical effort to harvest fish, not machinery. According to Gradman (1994: 105-6), older men draw on their life experiences to make sense of contemporary events.

Older men's critiques of technology and changes in the nature of fishing work are likely attempts to secure a consistent self-identity as hard workers and an elevated sense of the self over the "weakened" young. Older men fishers prided themselves in having fished in old age or even while injured or in ill health. Stories about refusing to get medical attention for an injury or condition so as not to lose a day's work were meant to demonstrate a certain individual toughness and strength. Younger fishers recited stories of "real fishermen" who fished regardless of declining health. It was their lifetime experience – their hard times past – that enabled older fishers to construct a critique of younger fishers' increasing reliance on new technologies and state assistance. Older men stressed hard work and traditional knowledge and developed an identity largely in the absence of complex and sophisticated technology, except for the watch and compass, motors, and new boat designs. Work was done by hand: "All by hand and some days you would be in some storms and wind but haul away at them, hold on. It's hard work, I tell you" (TEK-7, retired fisher, age 76).

While risk, danger, and hard labour were important elements in the social construction of the experience of small-boat fishers, especially those who did not acquire or could not afford new labour-saving and safety equipment, this construction served primarily to differentiate between old and young fishers. However, younger fishers also invested in the image of "man against nature," unmediated by technologies: "I always found it [fish] before ever anything [technology] came out. I think that that is the whole cause of the fisheries today, the technology is too much. . . . But the technology is gone out of control" (TEK-32, male fisher, age 50). In general, however, younger fishers recognized the existence of generational differences in the experience of fishing and were more likely than older fishers to bring attention to the long hours and the ability to work hard rather than the dangerous and laborious nature of the work itself: "I mean, you're never going to work as hard as they did years ago because, like I said, the technology is there, you know?" (NGP-1, fisher, age 35).

Part of the critique put forward by older fishers reflected a concern about the impact of the adoption of technologies on the generational and experiential knowledge and skills required, in their opinion, to be a "real fisherman." If technology mediated the relationship between fisher and nature, then fishers surely would be less able to acquire "a feel" for the winds and the sea necessary to set gear, and they would be less equipped to make assessments of danger than those who do not rely on such technologies. These losses were major concerns: "A lot of the young ones can't tell you nothing at all about that because they don't know" (TEK-34, retired male fisher, age 73). For older fishers, the ability to find fish and navigate with minimal aid was a valued skill. Thus, many believed that the adoption of navigational and fish-finding technologies had a deskilling effect, or at the very least, entailed an exchange of one set of skills for another. Yet traditional knowledge and local rules have become increasingly irrelevant as fishers are harvesting different species from and travelling farther distances than their fathers and grandfathers. Younger fishers also shared these concerns, expressed pride in knowing such traditional harvesting methods as handlining, and embraced the "traditional male fisher" model. In fact, some younger fishers claimed access to traditional ecological knowledge as a way to differentiate themselves from others and to justify their place in the future fishery. Consider the following remarks of a 46-year-old fisher:

> There was a lot of new people got into the fishery. They didn't know the grounds. . . . We would always use landmarks, hey? I didn't have a sounder. I could go on any rock out here in the bay, up the cape. I knowed hundreds of rock . . . after fishing with my father, you know? And when they got those Loran Cs, they got 'em in small boats, they come out on a rock we used and they punched it in, and then that's the way they get there. But if Loran C goes out they'd never find those rocks, you know? . . . There was a lot of young people started into the fishery and I don't think they understand it. . . . Those rocks, you've got to know the tide. . . . It's tricky work, you know? (TEK-3, male fisher, age 46)

There was a tendency among the older fisher respondents to report a distrust of the accuracy of modern equipment and an inclination not to use it or to rely on a combination of local ways of knowing and information from new technologies. The comment below demonstrates this skepticism and attachment to traditional methods: "What we got, we learned from our father – the marks, where to put our traps and where to drop all the rocks and everything. . . . We used to have a sounder the last years we was fishin' . . . I never put it on. I always go . . . and trap where we always fished to" (TEK-8, retired male fisher, age 69). To be sure, some younger men reiterated similar concerns regarding the unreliability of technology and the importance of traditional knowledge and methods:

> I still use the old traditional way, say, what we call rocks. . . . We're still using that. But I use a combination more or less because of what my father passed on to me. I mean, it's still there, it's just that, a combination is what I use now. With some, some of the *younger* fellers who never had that experience with their fathers and that so they uses this technology and that's it, you know? (NGP-1, male fisher, age 35; emphasis mine)

A couple of older retired male fishers reported having declined decades earlier offers from local fish-plant owners to skipper company-owned longliners when the vessels were first introduced in the region. Instead, they remained loyal to the small-boat fishery. These fishers were comfortable with the tools, boats, and technology they were already using. Some younger fishers read this behaviour as stubbornness: "Well, the way we see it, the older fellows weren't so good on the change part" (TEK-44, age 32).

Perhaps loyalty or stubbornness is best understood in terms of resistance to a sort of structural alienation. Fishers have always relied on "technology," however limited or minimal. They are proud of their traditional interactive and informed relationship with the tools of their trade: "Well, the way I got it sized up, we fellers, we're jack-of-all-trades. 'Cause we got to know how, well, basically know how to build a boat 'cause you got to keep it up. You can't afford to be paying people to be doing your maintenance. You got to know about engines and you got to know about electrical and twine" (NGP-5a, fisher, age 33). New unfamiliar technologies and computerized equipment have increasingly separated fishers from the tools of the trade. This is particularly relevant when it comes to repairs. Most fishers do not have the necessary knowledge, equipment, or skills required to deal with these technologies when they fail to work properly. In contrast to the way the "traditional" fishery functioned, fishers increasingly buy gear as well as services for repair work. In other words, fishers have had to relinquish some of their control over the work process with the adoption of modern technology because they lack skills necessary to use and especially fix the equipment. Yet, the purchasing of such unfamiliar technologies is interpreted by some as evidence of their professionalism.

Despite the obvious impact of technologies on the work experiences of fishers and the potential divisions, particularly along the lines of age, created by fishers in making sense of the impact, younger fishers did value traditional and experiential knowledge and practice. Younger fishers spoke with respect about older men who were "real fishermen" and sometimes with awe about the methods they used, the ecological knowledge they possessed, and the conditions under which they worked: "It is amazing what some of these old fishermen could do. . . . And they never had no sounders, no electronics to even find them shoals" (TEK-48, male fisher, age 42). In addition, traditional knowledge and practice continue to be relevant to fishers today. While risk, danger, and physicality *may* be minimized with technology, this depends heavily on a number of factors including appropriate training for and use of such equipment. Thus, traditional knowledge and practices remain important in constructing the ideal male fisher because they have functional value in day-to-day fishing and, historically speaking, survival has been contingent on such knowledge and practices. Disputing claims of dependence on technologies represents one way in which fishers are able to elevate the self and demonstrate toughness, self-sufficiency, and practical intelligence. At the same time, the need to fish further offshore to fill crab quotas has made traditional methods more irrelevant. Younger fishers have effectively integrated technologies into their work processes and subjectivities in ways that do not subdue or disorganize their perceptions of themselves as "traditional":

> You know, over the years everything, things change with technology and different kinds of equipment you try in the boat and that. But when it comes down to it, it's basically the same. . . . It [technology] doesn't really take much of the work. You got fish finder, you know, you go out on the rock, on the ground, you'll know the fish is there, but you still got to know where to go to take it, right? Because . . . you got to know how to work the wind and the tides. . . . The only thing like regard the labour part of it, ah, is the gurdy [winch]. (NGP-2, fisher, age 40)

## "THAT'S GOOD ENOUGH, BOYS, WE'RE DOING GOOD ENOUGH"

Some fishers noted that the introduction of more technology has changed the craft of fishing in another way – those with financial resources to acquire bigger boats and modern equipment have been given an edge in finding fish, which in turn has minimized the importance of traditional fishing methods and ways of learning, as well as family fishing secrets and traditional fishing berths. This is particularly important in the context of the enclosure of the commons via state restructuring. Many fishers were well aware of the potential link between increasing technology and incomes. The result at times was a generational conflict between the "being satisfied" ethic of the "traditional male fisher" model and the profit ethic of the "modern male fisher" model:

> We were saying to father, "Out of that $25 to 30,000 you are making, why not spend another $5 to 6,000 and perhaps you'll make $50 to 60,000 next year." But it seems he was content. But we weren't content. . . . And we were trying to explain to father, if we had a bigger boat we could take that, after three or four trips, we'd have an extra trip in that boat. "Nah, that's good enough, boys, we're doing good enough." (TEK-44, age 32)

Similarly, tensions between the egalitarian and the profit ethics were manifested in discussions about access to fishery resources, especially crab. There was support for the idea among small-boat fishers that scarce resources should be spread around: "Wouldn't you rather be satisfied with a little bit than with a big lot . . . this quota thing with this crab racket, that's the best thing since sliced bread as far as I'm concerned" (NGP-5a, male fisher). Small-boat fishers, who relied mostly on cod prior to 1992, insisted that they are entitled to crab quotas because they need to earn a living and there are few alternative lucrative species available: "Doesn't make sense. I've fished all my life, and I can't get a crab licence. I should be entitled to get any licence I want. I'm at it full time, that's all I ever done. . . . Everybody should be entitled to a little something anyway. . . . The bigger boats are making piles of money" (TEK-33, male fisher). Resource scarcity, though, has fostered conflict between full-time crab fishers and fishers who receive supplementary crab quotas.

The two full-time crab fishers argued that they are entitled to crab quotas because of their "traditional rights" to the species. They also raised concerns about the impact increased effort will have on the species if quotas are allocated to other fishers. Since 1992, fishers who had depended on cod for income have been demanding access to existing crab stocks and this has crab fishers worried: "What happened to the cod, happens to other species. . . . It was overfished. Now they are probably going to do the same thing with the crab. . . . When there was cod, there was only a few of us at the crab, it was enough to sustain it. But now, everybody wants to be at the crab. Well, the price is good. I can understand any man wanting to make a living" (TEK-10, crew on FPI crab vessel). "Traditional" crab fishers feared that because smaller boats make up the majority of harvesters, politicians will be pressured to meet the demand of small-boat, inshore fishers to have a piece of "their" quota. This concern leads to support for placing further limits on the fishery:

> Then the codfish started to get scarce. Then the other boats wanted part of *our* quota. And you can't blame them, everyone got to live . . . and the crab is the only thing that is left . . . the smaller boats, they got numbers . . . and the politicians got no choice only buckle to them. . . . Give it two or three years and we are not going to be allowed inside of fifty miles. They are driving the *traditional fleet* out of it altogether, and they are doing it through their politicians and lobbying the government in numbers. (TEK-18, full-time crab harvester; emphasis added)

Conflict was not restricted to the issue of one's "right" to access particular species. Choice of harvesting technologies was a particularly contentious issue. Fishers generally defended the gear types they employed as generating the best-quality fish and being the least ecologically invasive. Nonetheless, there appeared to be some consensus that hook and line and icing fish were best for quality. Hook and line, perceived by fishers as perhaps the most sustainable harvesting method, has been marginalized over the years with the adoption of new technologies and economic pressure to catch as much as possible. The particular versions of the future fishery espoused by fishers, unsurprisingly, are mediated by the gear types and boat size they employ. Small-boat fishers suggested the elimination of destructive draggers; trap fishers, the elimination of gill nets; and so forth. A trap fisher discussed the growing conflict between fishers who engage in such arguments. Rather than challenge state discourses regarding downsizing, he relegates the task of inclusion and exclusion to the state:

> Soon as I said that to a gill-netter, "You don't give a damn, you are all geared up for trap fishery, try to starve me," he said. "What if I said no more trap fishery, what would you say?" So stuff like that is something more than fishermen got to say it, because it only makes bad friends between us. That's what I find a lot of stuff going on here. They leave it to us to say, "You shouldn't get the crab licence, you shouldn't have this," and fight between ourselves. When the NCARP come out first, they called the Fishermen's Committees, which we recommended they do first, and we give them the list of fishermen in each community that are *true fishermen*, but then everybody was getting NCARP at the time. "Okay, boys, who shouldn't get it in the area now?" Why should I say to you, "*You* can't have that," or "*You* can't have that," after getting it. We just said, "No, you

> do it yourself." We had some fusses over that. (TEK-44, male fisher; emphasis added)

## "MY YOUNG FELLOW THERE . . . WHAT'S GOING TO HAPPEN TO HIM?"

It is evident that even where fishers ideologically resisted quotas, a limited fishery, and the entrepreneurial direction of the fishery, many felt forced to comply for economic and cultural reasons. However, fishers also made decisions about their work by considering their particular familial contexts. There is some evidence to suggest that having sons interested in pursuing the fishery provided reason not to move out of the fishery through early retirement schemes or not to relinquish licences through buy-back schemes:

> My young fellow there, that's all he ever done is in the boat with us. . . . And if something happened to me now, what's he going to do? He's only got his groundfish licence, and herring licence . . . I don't know if they are going to allow you to do that [pass on licences]. But if they don't, what's going to happen? . . . So, if I got to give up fishing, what's going to happen to him? That's something they [the government] should take into consideration, if it's truly handed down . . . I stops and thinks about this sometimes, what's going to happen to him, if he wants to carry on at it. It's pretty discouraging now for a young fellow. But he's sticking with me, doesn't want to see me go on by myself. Maybe if he wasn't around, maybe I would have considered selling it, or trying to get something for it. (TEK-43, male fisher)

Like other state policies, rules and procedures regarding intergenerational transfer – or, more accurately, patrilineal transfer – seemed unclear or uncertain to fishers. Could moving out of the fishery mean creating obstacles for sons to continue fishing? This uncertainty has generated fears among fishers regarding traditional inheritance mechanisms, in particular the ability to transfer fishing gear and property to sons:

> I got a young feller there, he left there last year in April month and he went away. I told him, I said, "X, it's up to yourself." I said, "You can stay home and go fishing with me or . . . you can pack up and leave and go somewhere else, because," I said, "there's no work around here anyway." But the way the government got everything here now, even if X did decide to stay home fishing with me, see, he'd have to go through hell and high water for to go and get everything I owns out there, right? (NGP-9a, male fisher)

According to the reform strategy implemented by DFO, new core members must replace exiting core members and licences in turn are distributed via the state as "replacement licences." These criteria appear to disrupt patrilineal inheritance patterns. However, the apprenticeship prerequisite and the ability of core fishers to name fishers for replacement licences might provide some benefit to those with links to the fishery. That said, the costs of the intergenerational transfer have increased substantially as the value of licences has skyrocketed. Because licences are not property, banks have been reluctant to accept them as collateral, forcing many to seek financing through processing companies – which opens up other spaces for company control (Canadian Council of Professional Fish Harvesters, 2001).

In light of such ambiguity, some fishers were reluctant to give up fishing because they hoped to ensure space on boats for their sons in the future fishery. Fisheries workers in general lamented what they understood as the gradual obliteration of a unique maritime culture and history. Fisheries workers talked about the disappearance of something tangible, the fish stocks, and something less tangible, fishing culture. The disruption of the culturally embedded intergenerational transfer of fishing know-how is likely to have a dramatic impact on who fishes, how people fish and under what conditions, and how that work is valued. Male and female fisheries workers shared concerns about their children's future and the impact of youth out-migration. According to respondents, fisheries collapses, and the state's responses to such collapses, strip their children of any real choice about remaining in outport communities and the fishery: "The way I look at it, what gives us the right to destroy everything in the ocean, make a desert out of the ocean, and leave nothing for our grandchildren, our great-grandchildren that are not even born? What gives us the right?" (TEK-48, male fisher). People – fishers and plant workers, women and men alike – are sad about the impact of recent events on young people's involvement in the fishery. However, it was clear that the loss for men and sons was perceived to be particularly detrimental: "I like to see the fishery open so the students can come out of university, get aboard the boat with their *father*, make a day's pay, and open up the fishery to everybody. I'm not sayin' now to teachers and people with jobs or anything like that, but every young *fellow* and every *man* that's involved with the fishery. Open it up, give them a chance to make a dollar" (TEK-8, retired male fisher; emphasis added). At the same time, fathers do not necessarily expect sons to stay in a failing industry or to incur huge debt trying: "I can't see why they would want to stay" (NGP-4a, male fisher).

Certainly, the precarious future of the fishery has placed an increased importance on formal education and training for youth. Many parents encouraged their children to leave their communities, to get an education, and to "do better" than they had by obtaining more secure and "easier" jobs outside the fishery. This sentiment, like fishers' views on many of the matters discussed thus far, is fraught with ambiguity. Below, a fisher mixes the "rational" language of the "modern male fisher" model with the "looking to the past" perspective and simultaneously wishes that his son escapes and participates in the fishery:

> *TEK*-53: My young fellow is in the boat two years now, and he is in university. . . . It is my *enterprise* and I should be able to give, that's why we started fishing in the beginning as a *family tradition*.
>
> *Interviewer*: You would like to be able to pass it on?
>
> *TEK*-53: *I would love to be able to do it. I hope he doesn't go at it.* I hope he's smart enough to be the engineer that he's going in for. But I got a brother got two kids, if one of them might not be smart enough, and the day that they can't fish because someone is telling them, "You can't get in the boat because you never traditionally fished before," I don't think that's fair either. (TEK-53, fisher, age 45, emphasis added)

Higher education, though, becomes less of a realistic possibility as available household money becomes scarcer with ecological declines. Neis (1993: 205) has pointed out that when money is scarce and land-based industries are geared towards men it is likely that opportunities for education will go to boys, not girls. At the same time, this withdrawal of sons' potential or real participation in the fishery means that fathers have little if anything to hand down to sons. Economic marginality has meant that inshore fishers, especially those who have been excluded from access to shellfish harvesting, may have little else to offer their sons except fishing knowledge and property. Fishers' fears about their ability to pass on their craft to sons take a particularly masculine form. As fewer young men become available to fish, a way of life will disappear, along with such valued practices as experiential learning and patrilineal inheritance. Interestingly, interviews with men plant workers indicated these same ideas. While not fishers themselves, men plant workers are connected to fishing through fathers, brothers, and the generalized masculine culture of fishing.

### "WE GOT SO MANY FISHERMEN THAT ARE *SUPPOSED* TO BE FISHERMEN"

Fishers not only established the criteria with which to define genuine fishers but also the criteria to exclude others from the fishery and state adjustment programs. Like the criteria used for inclusion, fishers' exclusionary criteria were created using contradictory "looking to the past" and "looking forward" discourses and were presented in ways that reflected themselves favourably. First, some fishers pointed to those who are somehow undeserving. Once again, fishers' arguments are often defensive in nature, pointing to unfair regulations that have excluded them from compensation packages or access to fisheries resources. While this argument is a critique of the state processes involved in determining eligibility, it simultaneously points to those workers who are perceived to have benefited from dishonesty: "The only reason I didn't qualify [for the financial compensation package] was because I was honest about it. . . . The boat that I fished in 1988, 1989, and 1990, everybody aboard that boat qualified except me. They fished the same times and days that I fished." Others supplied examples of "cheaters" – fishers who acquired false receipts from buyers in order to qualify for access to lump fish or wives of men fishers who falsely claimed fishing status to qualify for benefits. Through the use of comparison and example, men fishers defined "dishonest" fishers as non-genuine.

A second criterion for exclusion was the accusation of laziness. In this way, illegitimate fishers were described as those who have not fished since receipt of financial compensation because they prefer the easy route:

> I've seen some people that haven't went on the water since the moratorium, you know? . . . There's genuine fishermen and there's fishermen that got small ties to it as far as I'm concerned, right? You could, if you was out in Bonavista for any amount of time, you could tell who's the fishermen and who's the . . . yeah, who spends the time on the water and who doesn't, right? Some of those now is waiting for buy-backs, for to retire. (NGP-1, male fisher)

By fingering lazy fishers, they become the hard-working standard: "There are fellas I talk to who say that as long as the money keeps coming in from Ottawa, they are not going fishing no more, young people thirty-five and forty years old. 'What am I going fishing for if there is a cheque coming in every two weeks?' What a mentality to have.[4] It is only welfare anyway. I would love to be working all through the year, love it" (TEK-18, male fisher).

Third, men fishers were adamant about the need to eliminate "moonlighters," those who fished part-time during summer months and held regular paying jobs outside the fishery, such as schoolteachers: "Guys would not take their holidays until, they would wait and see when the fish are in and, 'I'm going to take my holidays now for a month.' Now they're out there competing with me, who's a full-time fishermen, trying to make a living. They're out there competing with me for that bit of fish" (NGP-2, male fisher). Meagre earnings often necessitated fishers to work outside the fishery to make ends meet, especially before the introduction of the welfare state. However, in the views of respondents, moonlighting is fundamentally different from a situation where a fisher needs to acquire work outside the fishery to make ends meet: "Well, years ago, that was okay, because some people didn't make much at the fishery. They had no other choice but to work at another job. In '87 I was fishing and I only made $1,600, so I had to find something to do to feed my family" (TEK-13, male fisher). The introduction of licensing in the post-World War II period generated much opposition from fishers because of the restrictions on moving in and out of the fishery in bad years. At the same time, fishers objected to an open fishery where people with full-time jobs outside the fishery could compete with full-time and "genuine" fishers:

> We got so many fishermen that are *supposed* to be fishermen, but they got trades in other fields, journeyman electricians, journeyman pipefitters. . . . The *professional fishermen* that never done anything else, that's the fellow that should be left there, not the fellows that are there for the easy money, "When the fish comes back, I'm going to go back fishing, and when times are tough, I am going to go to work and make $100,000 a year." (TEK-48, male fisher; emphasis added)

Those fishers adopting the "modern male fisher" model were especially preoccupied with establishing their occupation as a profession and limiting access. Cases where formal sea-related training enhanced the status of fishers support this view as well. This approach, however, does not easily coexist with the idea that all Newfoundlanders are entitled to fish for personal consumption – an idea I have associated with the "looking to the past" perspective.

Age was another exclusionary criterion. Heated debates among fisheries workers have occurred over the issues of age, retirement, and retraining: "[W]hat are they going to train me for, with so many young people out there, all kinds of trades . . . but this is one thing I never did agree with . . . to take a fellow out of the fishery and put them in trade school, when that was their livelihood, the fishery, and then there is another young fellow coming out of school, and can't get into trade school" (TEK-43, male fisher). Fisheries workers under the age of fifty-five were too young to take advantage of the early retirement option, yet often described themselves as being "too old" to retrain and locate work outside of the fishery.

There is an interesting dynamic at work here regarding how people subjectively defined age. Most fisheries workers believed that retraining would be more beneficial for "younger" men and women and were critical of young people who had failed to take advantage of retraining programs. Respondents indicated that it was easier for younger people to move and retrain. However, the "young" fisheries workers, especially men fishers, claimed that they were "too old" to retrain and that they had invested in the fishery too heavily, making removal an impossibility. For these men, age was certainly subjectively defined. It was determined through subjective assessments of attachment to the fishery and commitment to a way of life. After all, by age thirty, it is possible for men fishers to have accumulated up to fifteen or more years of fishing experience. The lack of other opportunities in their communities compounded this situation. For self-declared "young" fishers, there is no other choice but to stick with fishing, despite the obvious insecurities. Accepting the numbers argument and the need for a downsized fishery, some fishers limited exclusion to these eligible to retire, not "younger" fishers like themselves: "It's not room enough for everybody. But if they get so many out of it, and so many retire" (TEK-41, male fisher). Many younger and some older fishers suggested that fishers aged fifty-five and over should leave the fishery to relieve pressure on younger fishers to remove themselves. Some believed that older fishers should take advantage of the early retirement and buy-back options to give work to the younger generation: "People like fifty-five maybe they should . . . retire so that younger people can work, right?" (NGP-8, male fisher).

Not unexpectedly, there was some resistance to retirement on the part of older fishers, given how important fishing and "being a fisherman" are to them, personally and culturally: "I didn't give it up, they pushed me out. I went till I got drove out of it, and a lot more like me. That was all we could do" (TEK-12). Like younger men, older men constructed masculinity primarily in relation to fisheries work. Work provided a sense of belonging, an identity, and, of course, breadwinning opportunities, all of which are potentially lost in retirement. Gradman (1994: 106) writes, "As unemployment is painful, so the termination of work at retirement is no less painful." As one retired fisher explained: "Well, I'll tell you what it was like . . . it was just like a slap in the face" (NGP-3, retired male fisher, age 71). Unlike younger men, retired men fisheries workers could not anticipate a return to their work. Fishers retired since the 1992 moratorium spoke about a felt loss of control over their decision to leave work and often resented their lack of real choices in the context of moratoriums. Retraining was not an option. In fact, many suggested that they had planned to continue fishing if the cod moratorium had not been imposed: "No, I didn't retire at sixty-five. Maybe I wouldn't have been retired now.

Only for this retirement come up and the fishery closed down" (TEK-16, fisher, age 70).

The structure of fishing work has customarily conferred some control over the retirement "decision" to fishers. Fishing is not a job where retirement necessarily begins at age sixty-five, as is common in other occupations. One fisher respondent ceased fishing when the moratorium was declared in 1992 at age eighty. These men had expected to fish until they no longer wanted to or were no longer able, a time that was often based on subjective evaluations of physical health, just as their fathers and grandfathers had done. Older men have traditionally been able to remove themselves from fishing work due to the informal nature of the work group. Thus, being "forced out" is not taken easily. One retired male fisher (TEK-8) recounted a story about how his father "retired," illustrating how men fishers exercised some control over the timing of their departure from the fishery. Surely this measure of control aids in the transition from work to retirement: "When my father give it up, seventy-five, he give it up, so I took over then. He said, 'Get someone else,' he said, 'with you, that could help you out.' He said, 'I'm only in the way now. I'm not helping you at all, you know? You're only doing my work. Get a young man and take a lot of the work off you that I should be doing.'"

State-declared moratoriums have impeded this customary transitional process and stripped fishers of any real control over the shift from work to retirement. At the same time, labour-saving technologies on boats may mean that men feel they can continue to work despite aging bodies. Some older men tried to balance tensions between feeling forced to move out through state retirement schemes introduced since 1992 and feeling obligated to give younger men a chance at making a living fishing. Still others refused to retire: "I'm not finished yet. I look forward to it" (TEK-21, male fisher, age 65).[5] Existing research on men's experiences of retirement across class and job type (Gradman, 1994: 114-15; Solomon and Szwabo, 1994: 46) has suggested that masculine identification and attachment to work are linked to the status and nature of the occupation. Thus, because jobs characterized by low status and physical labour confer few rewards, men do not invest in such work as a site to display and express masculinity, which, in turn, facilitates their exit from work as the physical nature of their jobs becomes too onerous. In other words, the social and economic devaluation of such work enables men more easily to embrace retirement than is the case for other elderly men holding high-prestige jobs.

This line of thinking, however, fails to consider the complex cultural and subjective ways of defining status and masculinity. While fishing work may be stigmatized by the outside world and entail physical labour, local and subjective evaluations of this work are high, especially for men, and are embedded within a shared maritime history. Thus, we cannot expect the transition from active fisher to retirement will be an easy one psychologically, socially, or economically, especially in the particularly abrupt context in which retirement decisions had to be made due to the 1992 moratorium.

Skippers in this study who had retired attempted to maintain connections with male spaces and reported keeping track of their former crew's catches and other fisheries-related business. These connections were relatively easy to maintain because of the familial links and the relative proximity of homes. They continued to listen to the radio on shore to keep on top of fishing news. All of these attempts served to negate or weaken their increased association with a major "female" space – the home. This does not mean, however, that retired fishers were not aware of their new-found peripheral position in the fishery. Older fishers often responded to such marginalization by constructing difference between old and young fishers in order to save face and maintain a masculine sense of self. I have discussed some of these processes earlier in this book. For example, retired and older fishers drew on their early life experiences and their "hard times past" to differentiate themselves from today's youth, who they defined as lazy or "soft." Some accused younger fishers receiving compensation of faring better economically now than they ever did and linked state funding to the degradation of the work ethic.

Pre-existing divisions between older and younger fishers appear to have been exacerbated since the moratorium. I should note that some younger fishers also linked TAGS funding to a degraded work ethic among younger fishers. Some men who retired before the moratorium also prided themselves on having the "wisdom" to get out in time. In retrospect, given current ecological declines, fishers and others have construed decisions to retire as both timely and indicative of such wisdom. For example, one man (TEK-29, retired fisher, age 85) stated: "A man come after I give it up, he said, 'You must have known the time to give it up, because when you took your trap out of Capelin Cove the fish went out of it too.'" Of course, these constructions are divisive.

While it would be easy to polarize the "looking forward" and "looking to the past" discourses in opposition to one another, fishers used both discourses, sometimes separately, sometimes simultaneously, in pursuit of the same end goal – to meet their perceived material and ideological needs. The discourses highlight the model used by the state in policy-making, and these discourses in turn limit and constrain some possibilities and enable others. Male fishers "choose" to invest in particular ideologies and practices that confer a certain degree of advantage and support certain meanings, as well as a sense of continuity, even if this investment means adapting traditional work practices. At the same time, they seek to minimize economic and social insecurity. In their attempts to achieve stability, however, fishers sometimes develop contradictory responses. For example, it was not uncommon for an individual fisher to advance commitment to the traditional fishery and simultaneously be engaged in efforts to procure a larger boat and increase technological capacity. Contradictions such as these, however, become understandable when we interpret them as responses designed to hold on to fishing jobs and to maintain the ideological and actual spaces of maleness within their society.

## "THEY'RE CUTTING THE NEXT FELLOW'S THROAT"

As fishers make sense of state policies and their options using the "looking forward" and "looking to the past" discourses, they simultaneously facilitate and limit collective action. The exclusive discourses surrounding state allocation of financial resources and access to fisheries resources, internal disputes within the union, and the contradictory class consciousness of inshore fishers have divided fisheries workers, weakening collective action and the effectiveness of the oppositional possibilities and practices entailed in the "traditional male fisher" model. Small-boat inshore male fishers, in this study, saw themselves as fundamentally different from plant workers, crew on company vessels, and large-boat offshore fishers – even from female fishing partners. Further still, divisions among fishers existed, for example, between old and young and between trap and gill-net fishers. At the community level fisheries people are pitted against non-fisheries people.

The targeted allocation of financial compensation packages and early retirement options created divisions among fisheries workers (Sinclair, 1999: 336), who felt a sense of unfairness and exclusion in the allocation process. In fact, divisions extend beyond fisheries workers. Many local people who work outside the fishery have been extremely critical of financial compensation provided to displaced fisheries workers. Financial compensation packages were marked by different qualifying periods for different groups of fisheries workers, largely based on state assessments of economic attachment to northern cod. It should not be surprising that such state assessments have served to disadvantage women and plant workers more than men and fishers (Neis, 1993; Williams, 1996). Yet men fishers rarely recognized this relatively privileged position.

Targeted programs and different qualifying periods have fostered tensions among the parties affected and have muted collective support. One woman (NGP-9b) told me about an organized protest by fisheries workers in the summer of 1998 aimed at demonstrating dissatisfaction with upcoming terminations of financial compensation for particular individuals. However, recalling that no such protest had been organized when *she* was cut from the package, she defended her lack of support: "The thing that gets me is why now, when there's only a few left [on TAGS] that's going to be good till '99? . . . Why protest for those few? What about the ones like myself that got cut off last year and a few months ago?" Such lack of support for organized protests and other union-related activity should not be taken as evidence of a lack of public protest by fisheries workers since 1992. Various issues of *The Union Advocate*, the official quarterly of the Fish, Food and Allied Workers Union, from 1992 onward include pictures and articles about organized opposition to state policies and programs. However, such protests have tended to be specific in nature. Protests tended to represent the interests of a specific group of fisheries workers, such as crab harvesters or processors at a particular plant at a particular time, and were directed at specific state initiatives that were disadvantageous to that group. Where fishers have organized collectively, the protests tend to be issue-based.

The union, FFAW, facilitated collective action but was also a source of division among fisheries workers. The union's members include plant workers and fishers from both the inshore and offshore sectors. There was support for the union: "The union has been an important factor in the Newfoundland fishery. . . . It has been an important factor in the price of fish for Newfoundland" (TEK-22). However, many, including some supporters, believed that wide union membership acted as an impediment to effective representation of the interests of *their* particular group, for example, processing workers, inshore fishers, or company crew. Wide membership appeared acutely problematic for inshore fishers, who regarded themselves as independent: "I don't know how they [the union] all hammers it out. I mean, especially here you got the same union involved with the fish plants and we're on our own, you know? We're not employed by no fish plant, although our cheques comes from a fish plant" (NGP-5a, male fisher).

Concerns about wide representation are not new, but they gain significance when resources are scarce and the union must fight on behalf of competing interests: ". . . now they [the union] tries to help the full-timers, they're cutting the next fellow's throat. Then he helps the other fellow, he's cutting the full-timers' throat. So I can't see how they could do it anyway . . . they got to hurt one to help the other" (NGP-4a, male fisher). Given the particular social and economic climate, respondents consistently suggested that somebody else or some other group was getting the better deal. These divisions become glaringly obvious in discussions about access to resources. The allocation of crab quotas is a particularly touchy subject because moratoriums on groundfish and other declining stocks mean that access to crab is the only real means to making a living:

> [I]t's our sector, small-boat crab, and there's the supplementary and then there's the full-time, and each one is blaming the other one, blaming the union for doing more for them. . . . I suppose where they're representing three or four different sectors, I guess you can't do everything for everyone. . . . I think the union could have fought harder for us fellows . . . for the small boats, that is, right? (NGP-1, fisher)

Many fisheries workers complained that the union represents the interests of "the same old crowd" (NGP-11, male plant worker) all the time. One male fisher (TEK-27), an active member of the union, agreed that nepotism is a problem: "What I've observed over the years, especially on those boards and committees, the greater majority is just there for themselves. . . . But that's no reflection on the union, because how do the Earle McCurdys and the Reg Ansteys,[6] or the people from Toronto, know the difference?"

Social cohesion has been disrupted as local communities fight over resources. During the interviews, a number of respondents voiced concerns over competition between communities for scarce marine resources. One particular source of contention at that time related to FPI's rumoured consideration of moving crab processing from the plant in Bonavista, an older processing facility, to the Port Union factory, a modern and large plant (Sinclair, 1999: 337). According to Bonavista residents, "their" plant had traditionally processed crab and, thus, should continue to do so. Port Union and Catalina residents, on the other hand, suggested that because "their" facility is equipped with modern technology and offers a larger production space, it would make better sense to transfer crab processing from Bonavista to Port Union. The impact that such a move would have on who gets hired was a major concern for men and women processing workers, as well as for fishers, who preferred to supply a plant where family members work. Because the site of processing impacts local economic activity, local business people were also concerned. In the end, crab processing remained in Bonavista and the Port Union plant was transformed to process shrimp.

Another source of local tension surrounded who should have access to the little existing processing work at the Bonavista crab plant. Formally, this work was distributed among the most senior workers in the area, most of whom are men because of the gendered nature of seniority lists. This meant that workers from the Port Union plant were also getting occasional work at the Bonavista plant. Former Port Union workers accused the Bonavista shop stewards and supervisors of unfairly marginalizing "the Catalina crowd" (NGP-16b, female plant worker). A number of Catalina and Port Union residents suggested that the "Bonavista crowd" (NGP-16b, female plant worker) did not want outsiders taking away work in "their" plant. These conflicts are not new and are rooted in historical community antagonism. However, such conflicts are also linked to economic vulnerability.

Individuals tried to secure a place for themselves and "their" plant by defining themselves as part of a larger workforce that was "the best around." In this way, workers took part in objectifying themselves as valuable, if not irreplaceable, commodities. The work experience of plant workers has long been plagued by threats of plant closures. However, the issue of job security is exacerbated with moratoriums, making it increasingly important for workers to appear loyal and compliant.[7] Workers are motivated to appear compliant especially when they need their jobs and want to work for pay. Women and men alike preferred to "earn" income rather than receive government money, which they interpreted as a handout. Men and women respondents reported a preference to receive unemployment insurance benefits rather than NCARP or TAGS. Unemployment insurance benefits tended to be perceived differently from welfare and NCARP or TAGS money because of its attachment to paid work (Ommer, 1998): "If a job comes up here, if there's one job we got twenty people looking for it, right? So people don't want to be on TAGS. People rather be working" (NGP-10, male plant worker).

As families face increasing uncertainty and economic pressure within the context of fisheries crisis and moratoriums, individualism – characteristic of the traditional Newfoundland household-based fishery (Porter, 1993: 49) – has been the prevailing response. Opportunities for fishers, fisheries workers, and even members of fisheries-dependent communities to organize collectively against the increased penetration of capital have not materialized. The specific reactions described here may reflect the specific timing of the two sets of interviews. However, to date most organized efforts of resistance remain specific, local, and even populist.

## SUMMARY

A major shift is occurring in Canadian fisheries management. The state is in the process of dismantling its extensive fisheries regulatory regime and transferring the responsibilities for management onto participants in the industry. This corporate model is evident in the state's efforts to enclose what had been the commons via boat quotas and to limit participation through the core/non-core classification system and support for professionalization efforts that correspond with its direction. At the level of policy, the state is establishing a "transnational business masculinity" (Connell, 1998). Individual male fishers have responded with support for and resistance to this direction in their efforts to claim a place in the future downsized fishery. Drawing on two competing local discourses – "looking to the past" and "looking forward" – and accepting the "too many fishers, too few fish" argument, fishers argued for further enclosure, increased state management, or both. Either way, fishers developed criteria for their personal inclusion in a future fishery that tended to reflect their positions. Some criteria are clearly rooted in the "traditional male fisher" model, as in the case where fishers emphasized long working careers and historical dependency. Others reflect an investment, however shaky, in the "modern male fisher" model. Psychological and financial investments in "going bigger" and new technologies illustrate this model.

While some fishers clearly aligned themselves with one side or the other, most seemed to straddle the two sides uneasily. This reflects, in part, confusion on fishers' part over the state's plans regarding the future fishery and how resources will be distributed, and over what they perceived to be ever-changing rules. It also reflects their unfamiliarity with neo-liberal economics – both the jargon and the philosophy. This confusion proved particularly problematic for fishers who wish to pass on the fishing tradition to sons. Fishers concurrently formulated exclusionary criteria using the two competing discourses. According to fishers, men with a "dishonest" or "lazy" character, moonlighters, and older fishers qualifying for retirement should be excluded in the future fishery. Again, individual fishers emphasized one, all, or any combination of these criteria for exclusion.

These discourses and models are so important because they are about both access to and the distribution and control of ideas and resources. Kaufman (1994: 145) has suggested that "[e]ach subgroup, based on race, class, sexual orientation, or whatever, defines manhood in ways that conform to the economic and social possibilities of that group."Available material resources and discourses shape people's daily lives, identity formation, and how they make sense of the world, which in turn inhibit or enable social action. Motivated by their fear of exclusion, collectively and individually, fishers expressed conflicting, ambivalent, and divisive responses. This fear was not limited to fishers. Rather, it spilled over into wider social relations as communities and workers fought for access to scarce resources. The contradictory class position of inshore fishers appears to be enabling, constraining, and mediating changes in local configurations of practice and ways of thinking.

The "modern male fisher" model is challenging the local hegemonic version of masculinity, the "traditional male fisher." This challenge is particularly significant because it comes from both outside (i.e., the state) and inside the local culture. The success of this challenge to the more subversive "traditional" model is undoubtedly linked to fishers' material conditions. Declining stocks and incomes will likely boost support for individual quotas – to secure access to fisheries wealth – and individual transferable quotas – to enable fishers to sell and exit a declining industry, and these developments will undermine commitment to an open fishery based on egalitarianism. Fishers' decisions will also likely affect sons' decisions regarding entry into the fishery, the makeup of the next generation of fishers in Newfoundland, and the conditions under which they work. One thing seems certain, however; the fishery will remain predominantly male.

# *The Patriarchal Dividend** 6

## WOMEN'S PLACE IN THE FISHERY

The effects on the lives of fisheries-dependent peoples of the collapsed stocks, the state's responses to these collapses, and the more general trends of government withdrawal of social spending and its privileging of the free market are mediated by gendered and hierarchical divisions of labour in the fishery, within households, and in communities. Restructuring initiatives in the Atlantic fisheries have primarily targeted an assumed male breadwinner. Assumptions in policies and programs about gender divisions of labour shape access to fisheries-related wealth, resources, and jobs, as well as spaces in which to create meaning and identity. Male fishers have responded to state initiatives designed to deal with fisheries crises by looking to the past and investing in the "traditional male fisher" model or by looking forward and investing in the "modern male fisher" model. These models – and the resources that accompany them – are not readily available to women. A history of patriarchal local customs and ideologies and state policies have reinforced the idea that fishing is a male domain and maintained sexual divisions of labour and male control of the technical means of production and fisheries wealth. Men accrue the patriarchal dividend from the outside world, including access to state money and programs, by virtue of being recognized as the economic heads of households, and they benefit from the material relations around reproductive and care work.

The sexual divisions of labour in the traditional household-based salt fishery and the modern fresh/frozen fish industry can be described in terms of the land and sea divide (Porter, 1993: 82-3; Wright, 1995: 129). In general, fishing is and has been men's work.

Women, on the other hand, tend to be relegated to processing and the household. However, to make such a rigid dichotomy between land and sea, between women and men, would be simplistic. Contemporary women do fish and historically women have been involved in harvesting. Women in Newfoundland were hired to cook and perform other duties in the Labrador "floater" operations[1] (Andersen and Wadel, 1972b: 141). Szala (1977, in Nadel-Klein and Davis, 1988: 29) noted that women from the southwest coast of Newfoundland caught fish for consumption when husbands were away. Women have also been responsible for other fisheries-related work such as keeping records of fish landings and receipts.[2]

A history of poverty and exploitation by merchants, and later by processing companies, has necessitated the economic participation of all household members to survive. Consequently, men and women are likely to have crossed the land/sea and work/home divides to ensure that the required work got done (Cadigan, 1995: 52; Neis, 1993: 190-4; Porter, 1993: 19, 44, 91). In addition, male absence from the household has served to limit direct male control over women and has given women responsibility for and authority over land-based activities both historically and more recently (Andersen and Wadel, 1972b: 142). The contradictions in the local gender order no doubt facilitated the existence of good marriages and independent women. A number of authors have recorded the existence of stable, reciprocal, and gentle marital relationships, historically and in more recent times (see Davis, 1983, 1988, 1993; Porter, 1993; Stiles, 1972; Szala, 1978). Porter (1993: 49-53) suggests that the historically accepted sexual geography and divisions of labour provided women with the "physical and ideological space" necessary to construct arenas of independence and autonomy, and also argues that the wives of boat owners actively worked to secure sexual boundaries to ensure female control over women's spheres. In a context of poverty, women would have been less concerned about securing direct access to fisheries resources and property.

Andersen and Wadel (1972b: 142-3) suggest that using the land/sea divide to describe gender relations in Newfoundland is problematic for another reason, namely, while the sea is a male sphere, the land is not exclusively a female domain. Patrilineal inheritance of property and knowledge, patrilocal residence, male preference, and local and state-produced patriarchal ideologies have not only denied women access to fishery resources but have shaped the sort of work women do on land (Neis, 1993: 191-2; Silk, 1995: 265-6; Wright, 1995). This means that women's authority in the household or in other female spheres has likely been limited both in and outside these spaces, especially in a context where men dominate public economic and political relations (Porter, 1993: 49). In addition, community spaces associated with fishing, including boats, wharves, and stages, are regarded as male spaces (Silk, 1995: 265). Finally, the land/sea divide is problematic because, despite women's historical participation in processing work, this work is not considered "feminine." The sexual division of labour within processing plants has positioned women and men in different jobs, perpetuating the idea of sex difference and its connections to space, activity, and physicality, without conferring "femininity" to women performing these jobs.

It has also been argued that the historical recognition by locals of women's fisheries work has been undermined in the post-World War II period. The modernization of Newfoundland's fisheries included processing plants and a cash economy, as well as an ideology of separate spheres for men and women (McCay, 1988: 115; Wright, 1995: 141-2). This ideology was reflected in state policies aimed at transforming women salt-fish makers into unpaid housewives and consumers. Economic reality, however, necessitated that women work as "secondary" earners in fish plants, where they were paid less than men either legally or later via sexual divisions of labour and patriarchal evaluations of skill (Neis, 1993; Robinson, 1995: 170-3). Of course, this work was never really secondary, given its necessary role in the reproduction of the household. Porter (1993: 126) writes: "When all members contributed economically to the household, but none of them were paid, it was easier to see women's contribution for what it was – equally important to men's." Women's subordination, then, has been reflected in and reproduced by the ideologies and practices that make women responsible for domestic and caring work, the undervaluation of this unpaid economic work, and the view that women are secondary earners – all of which mediate the impact of restructuring on women and their abilities to cope.

## "I GOT NO HOPE TO GETTING BACK TO THE PLANT"

Like men, women experienced the negative impact of moratoriums through a loss of work and income and a sense of uncertainty about their futures. However, strategies adopted by women to deal with the crisis are shaped by available opportunities, which differ from those available to men. Women have been arguably more disadvantaged than men. This argument, though, has been challenged by others including Davis (1993). However, her work fails to conceptualize local gender relations as power relations and she consequently perceives men as victims, ignoring women's subordination. Her argument that men are more negatively impacted by the state-imposed moratorium than women ignores the real economic and social marginality experienced by women within the fishing industry. In addition, while men might perceive women to be competitors for scarce resources, as Davis contends, this does not accurately reflect the reality in terms of the allocation of state funding and possible employment opportunities in outport communities (see Robinson, 1995). Women have had fewer opportunities than men fishers and plant workers in maintaining paid fisheries work since the moratorium. Women worked predominantly in the processing sector, which was hit particularly hard by moratoriums and stock collapses and they also occupied the more vulnerable processing positions there (Neis et al., 2001: 40).

Women plant workers who had been employed at either plant in this study tended to experience downtime and layoffs earlier than men plant workers because of their places on seniority lists – gendered by design in their failure to recognize that women's reproductive work interrupts paid work, which has meant that women take up processing work later in life and have had less time to build seniority (Power, 1997). According to some respondents, when processing shifted almost entirely to snow crab in the Bonavista plant, some male plant workers moved into jobs at crab processing, which were traditionally women's jobs, displacing less senior female workers. Also, men were hired at the Bonavista plant for renovations and repairs. It appears that the "nature" of these jobs precluded the hiring of women.

Although most fisheries workers received state-funded financial adjustment benefits for varying and limited periods, incomes for many were drastically reduced. Displaced women fishers and plant workers have suffered disproportionately, receiving fewer and smaller adjustment payments (Williams, 1996: 23-6). Recent policy changes have restricted both fishers' and plant workers' access to unemployment insurance (Neis and Williams, 1997: 52-3), with the result that women's paid work outside the fishery is even more important to household income. In some cases, women have become the sole breadwinners. Although women's contribution has increased proportionately, sex-segregated labour markets, the wage gap, and the need for part-time work (to accommodate child-care duties) mean that women still bring lower incomes to the household than do men. Women's lower wages and earnings present a serious challenge to households that are relying increasingly on them.

In addition to the obstacles women faced in terms of maintaining employment in the fishery during the moratorium period, women's opportunities to retrain were severely constrained (Human Resources Development Canada, 1998: 58; Neis, 2001: 32-3). Women, like men, were very critical of educational and retraining opportunities. Women generally agreed with the widespread view among men that the state's intention behind such "opportunities" was to remove people from the fishery and rural communities. However, the failure of women to participate in retraining can be explained only partially in terms of resistance to removal from the fishery and their communities. A closer look illustrates that women's lack of participation in retraining is as much about their lack of opportunity as it is resistance. A common complaint was that available educational programs were not "appropriate" for women. Current research on displaced women fisheries workers in Newfoundland has demonstrated that women are less likely than men to find alternative work because adjustment and retraining have been focused on men. Robinson (1995: 170-3) has argued that women have been systematically disadvantaged through the allocation of funding and opportunities because of a pervading assumption that women are unskilled, secondary earners. In addition, women respondents did not take advantage of retraining due to practical concerns of time, energy, and home responsibilities. The tongue-in-cheek banter below between a wife and husband (both had worked at the local fish plant) highlights some of the difficulties women faced in terms of retraining opportunities and leaving their communities:

> NGP: What do you think about retraining programs and other educational programs being offered?
>
> *Wife* (NGP-13b): Well, I thought it was wonderful. I thought it was really a good opportunity. I must say when the moratorium came on and they came up with those programs, it was ideal for younger people. But I couldn't see it, I honestly couldn't see it for X (her husband). I couldn't even at that time see it for me. Well, I was in my forties. First I'd have to upgrade to grade twelve . . . and I didn't believe in just training for something you didn't want to do, just to train because to say, "Well, I got to train, well, if not I'm going to lose my TAGS." I think that was the biggest mistake people made. If I wanted to do it I would like to do something that I'd want to do, you know? And I did talk about going and X (her husband) said, "Well, if you want to do it, do it, but I'm not going." You know? So here.
>
> *Husband* (NGP-13a): Did I say that?
>
> *Wife*: Yes, you said that. And so what was I going to do, you know? Go and leave my kids?

> *Husband*: You could divorce me.
>
> *Wife*: Well.
>
> *Husband*: Thought about it, never had money enough, did you? (laughs)
>
> *Wife*: If I knew TAGS was going to last this long maybe I'd had done it but . . . (laughs)

While husbands, especially displaced male plant workers, are free, at least ideologically, to leave their families and communities to take part in retraining programs or find alternative work outside of their communities, women have less opportunity to do so. A lack of money may have been a concern for both women and men, but women had the ideology of motherhood constraining them. According to local ideology, women who left their children were "bad" mothers. Women also reported that where husbands decided to relocate, wives and families followed, but that it was unlikely that husbands would follow wives: "I find if the man is going, nine times out of ten the whole family go. This is what we're finding is happening here now" (NGP-17, female plant worker). Women's experiences of displacement, then, are different from men's. Despite unemployment and few job opportunities, women remain in these communities due to familial ties and commitments. The obstacles that women face in terms of maintaining fisheries work and retraining are linked to the gendered nature of fisheries work. Women were not inclined to express the same sense of missing fisheries work as men did. They tended to miss their paycheques rather than the actual work. Women complained about a felt loss of independence that accompanied displacement largely connected to no longer earning "their own money" at the local fish plant. Women interpreted being cut from TAGS in a similar fashion: "Well, it's a big cut you know, a really big cut, but, I mean, it was worse for some people. Like it's not too bad for me. I mean, my husband is working. But still I can't see when it do cut off not getting a cheque of my own" (NGP-18, female plant worker). While women respondents understood their independence, at least to some degree, as contingent on the wages they brought home, these wages were hardly supplementary. Rather, in keeping with gender roles, women's wages were used to purchase much-needed household and child-related goods, but also to pay bills and purchase groceries. Women tended to perceive their paycheques as demonstrating their commitment to household survival and their children's well-being and, as such, they considered their paid work as an extension of domestic responsibilities.

While women did not miss plant work in the same ways men did, it was clear that many women missed the social and psychological benefits of their paid work. Women often described plant work as a "break" from their domestic and child-care routines and some suggested that wages enhanced their status. Generally speaking, the women respondents, unlike the men, did not invest a sense of self in their paid fisheries work. There are differences in terms of female identification to the fishery along the lines of age however (see Nadel-Klein and Davis, 1988; Neis, 1993; Porter, 1993; Power, 1997). Older women are more likely than younger women to identify with the fishery because of their direct participation in the production of salt fish. The post-Confederation female experience of the fishery, however, has been more ambiguous and tension-ridden than men's. This female experience can be explained in terms of sexual divisions of labour mediated through changes in the relations of production and the spread of a "modern" gender ideology. Contemporary women processing workers are distanced from the fishery because of their positions in factories and capitalist organizations. While contemporary women plant workers may not invest their sense of self or identity in paid fisheries work in the same ways that men do, they do attempt to create meaning and derive dignity from their work. For example, many women expressed pride in doing their jobs well. It was not uncommon for women respondents to suggest that the sexual divisions of labour in processing reflected the ability of women to do work that men could not. Such claims may reflect women's attempts to make sense of their unsatisfying work experiences by "rationalising oppression, making a virtue out of necessity" (Pollert, 1981: 122).

While women's identities are less bound up with the fishery than men's, this does not mean that they are any less dependent, economically and otherwise, on fisheries-related work. Women respondents, like men, were grateful to have had jobs, which paid relatively decent wage rates, in communities where high unemployment rates existed. Clearly, women's wages from fisheries work and unemployment insurance benefits helped ensure the survival of the family. Plant work and fishing, at least, allowed some flexibility in terms of shifts, enabling women to juggle both paid and unpaid work: "I suppose where I'm a woman, more or less where we[3] got work to do, it's, night shift would be better for us" (NGP-14b, female plant worker). However, this simply added up to the double day. In addition, fish factory work provided good incomes for women relative to other available "female" jobs, such as clerks in local shops or wait staff in bars and restaurants. The combination of income from seasonal plant work and unemployment insurance benefits was an ideal way for some women to juggle child-care and household responsibilities and still bring in a badly needed second income: "I'm not a person who really believes in unemployment insurance, after you get so many stamps you quit and stuff like that, but I think for a working mother it was ideal" (NGP-13b, female plant worker).

Recognizing their limited opportunities for paid work in these communities, women respondents often expressed a willingness to move away. This willingness likely reflects their perceived economic possibilities in post-moratorium communities. However, women were also fully aware of the commitment on the part of their male partners to remain in the fishery and fishing communities, as well as the obstacles women, and men, faced in terms of relocating, which muted this willingness to some degree: "Well, if I got to I would, yes. Now I would move, but like X (her husband) said, he don't think he would. . . . Like, I got no hope to getting back to the plant at all" (NGP-14b, female plant worker, age 40). Similarly, another female plant worker explained (NGP-17): "No, I'm flexible, right? I could move on to something else. Like I could move from here, and it wouldn't bother me."

Perhaps the most obvious reason for the differing degrees of perceived investment of self in work between men fishers and women plant workers is the nature of their respective jobs. Women's waged work in the fishery as fish-plant workers does not entail the same work autonomy or cultural prestige that men fishers experienced. Women assessed plant work as simply a way of making money; it was a job like any other manual job. Men fishers, on the other hand, assessed fishing work as liberating and as providing a place where they could accrue cultural significance and advantage in a world that affords them few other ways to access economic resources. Processing work takes place in an environment often characterized by poor conditions, heavy monitoring, rising standards, and performance requirements, especially with the introduction of individual work stations and incentive systems, and with threats of expendability with the introduction of machines (Fishery Research Group, 1986: 42).

While male plant workers mentioned many of these complaints, women's experiences as commodified workers were exacerbated by the fact that they are women. The organization of fish processing through sex segregation has a differential impact on men and women (Fishery Research Group, 1986; Neis and Williams, 1993; Power, 1997; Rowe, 1991).[4] Women tend to be supervised by men, concentrated in low-paying, monotonous production-floor jobs and clerical work, and subjected to work processes associated with a host of work-related injuries. Women also experience sexual harassment from supervisors. Men's jobs at fish plants tend to be more varied. For example, cutting, jobs in discharge and service, as well as in the holding room and freezing departments, are filled by men. Packing jobs, however, seem to be women's work. Not only is women's plant work more vulnerable, monotonous, and heavily supervised than men's plant work, it is also undervalued. Women's factory work is undervalued because it is linked to "natural" female abilities – as opposed to skill – and women's status as supposed secondary wage earners. Just as the dominant, patriarchal work paradigm holds mental above manual work and reflects gender and class relations, so, too, "women's manual work, because it is women's and because it is manual, is doubly under-rated" (Pollert, 1981: 65). This argument is certainly supported in the existing literature on processing work in Newfoundland (see Fishery Research Group, 1986; Rowe, 1991; Wright, 1995).

The sexual division of labour in processing plants also mediates men and women plant workers' resistance to moving from their communities and subjective investment to the fishery. In general, men plant workers, like men fishers, opposed leaving their communities. Men's experiences of work in the factory, as well as the time spent outside of work, differ drastically from women's. Men's plant work is arguably less restrictive, supervised, and monotonous than women's. In addition, it is assigned more value than women's work and regarded as "men's work" proper, considered too difficult for women to do. Also, men's free time is spent, at least to some extent, living the "traditional" life. Men plant workers participate in such culturally valued male activities as hunting and continue to occupy male spaces in the community. Outside of work, men plant workers are free to carry on the traditional male life. Women have another job to do – their domestic unpaid labour.

The difference in the work experiences and productive positions of women plant workers and men fishers explains, at least in part, how men and women understood their relationship to the fishery. Indeed, women who fished voiced substantially more resistance to perceived state-directed attempts to remove them from their communities than women plant workers: "Well, I'm not, most definitely not, uprooting. . . . Where am I going? . . . I'm just as well off here as what I am if I got ready and went to the mainland. . . . And no way in the world would I sell something that I worked for all of my life and now when we should have comfort. Our children are grown" (NGP-9b, female fisher). The women fisher respondents preferred fishing to working in the local fish plant for many of the same reasons as men, despite the difficulties entailed in harvesting: "But she [the boat] had no house or nothing on her, just outboard motor in the open boat. You go up the shore with water coming in your face. . . . You'd go out in the morning, you haul and haul. Believe me, it wasn't easy. But I still preferred it to the fish plant" (TEK-19b, female fisher and plant worker). Another female fisher (NGP-9b, female fisher) stated:

> I worked in the plant for twelve year and I don't know, it just got on my nerves. So in '86 myself and X (her husband) went handlining. That was the only thing that I did that year was the handlining. And I liked it. I really enjoyed it because you're out in the fresh air and everything. So then in '87 we did the same thing, well, X got his lobster licence in '86, . . . so in '87 we started at the lobsters. So then I used to go out with him and haul the pots, then we go handlining in the fall. So I finished at the plant then, I didn't bother to get back because I was making enough during the summer to draw my unemployment in the winter.

This observation might lead us to believe that we can reduce differences in levels of resistance and subjective attachment to the fishery to position within productive relations. However, men's and women's experiences of displacement are also different because of how gender relations interact with household and unpaid work. Women's attachment to the fishery has been regarded as secondary to their domestic and caring responsibilities. Indeed, women's relegation to the domestic sphere enabled women to interpret their participation in paid fisheries work as an extension of their domestic responsibilities and, thus, secondary. This is, of course, an argument that cannot be used to exclude men plant workers from the masculine culture of the fishery. Relegating female responsibility primarily to the domestic sphere also allowed men and women to explain away women's entry into the processing or harvesting sectors as a temporary or secondary household strategy. In this way, women's attachment to the fishery could be defined as loose, and qualitatively different from the vocational commitment to and sense of identity derived from the fishery experienced by men. Women interpreted their fisheries work as an extension of their domestic and caring responsibilities and were committed to do whatever it took to make their lives and the lives of their children better. Most women plant workers said they "chose" work at the plant because they needed "extra money." The reality is, however, that working-class and fisheries-dependent households depended on women's fishery incomes and unpaid work to survive and women's fishery-related work, paid and unpaid, was often not so temporary. Many of the female respondents spent much of their working lives in the plant, whether they had intended to or not.

Gendered household and subsistence arrangements also shape men's free time. Free from domestic responsibilities, men were able to live life according to the "traditional male fisher" model. Male plant workers, like fishers, continued to participate in such culturally valued male activities as hunting and to occupy male spaces, including those connected to the fishery such as private fishing premises and community-based wharves, public spaces such as town halls, pubs, and sites designated for sport, and the surrounding "natural" environment. Men were able to participate in these activities and to have such freedom only because women's unpaid domestic and caring work freed them to do so. In their local villages men's language, the clothes they wear, and the work they do make sense and are valued. Outside the fishery and outside their communities, these things would be lost or at least make little sense. Local communities are safe havens for these men. For men, then, leaving the fishery and their communities would mean losing male privilege that is embedded within culturally ascribed work and social relations. The situation for women plant workers (and women fishers), on the other hand, is drastically different. As Porter (1993: 88) has explained, contemporary women are not able to take part in the so-called "traditional" life like men. Instead, the disappearance of the role of the fisher's wife since the decline of the salt fishery, the dissemination of a dominant ideology that places women in the home, a historically rigid sexual division of labour, and the introduction of modern conveniences have created new roles for women (see Nadel-Klein and Davis, 1988; Neis, 1993; Porter, 1993; Power, 1997). Contemporary women living in fisheries-dependent households find themselves responsible for unpaid domestic work in addition to their paid labour. Yet this unpaid work tends to be unrecognized and undervalued within the local culture and at the level of policy-making.

## "WE NEEDED THE EXTRA INCOME"

The gender-neutral discourse surrounding privatization, professionalization, and the allocation of quotas is not gender-neutral in its impacts. The criteria used to determine core status – heading an enterprise, holding key licences, and demonstrating attachment to and dependency on the fishery – will make it difficult for women to enter independently and will reproduce women's general subordinate status in the harvesting sector. As in other North Atlantic fishing economies (Munk-Madsen, 1998: 234; Skaptadottir and Proppe, 2000), when the right to access fisheries resources depends on ownership of fishing capital, the patriarchal dividend is upheld or created where that capital tends to be the property of men and where there are no formal mechanisms whereby women can exercise a say in how the property or rights are used. Munk-Madsen (1998: 235) writes of the Norwegian fishery: "Men have received their fishing rights by building up fishing capital through their exploitation of common property resources and government subsidies, well supported by women's unwaged labour." Investment in fishing boats and technology affects all members of fishing households, not simply fisher husbands. Few Newfoundland women are eligible because they do not own any or at least any key licences, and they have shorter or interrupted fishing careers and thus less total and annual fisheries income. For many women fishers, the criteria represent a Catch-22 – access to fishing licences is limited to core fishers, yet to be a core fisher one must hold key licences.

According to Neis et al. (2001: 97-100), women tend to have limited fisheries training – another criterion for progress in the industry – which means they are further disadvantaged. The cost of upgrading one's skills or accreditation through training may be crippling if a household has multiple members in the harvesting sector. Most women fishers participate in the fishery via husbands or male relatives. If advancement in the industry requires formal training and a household cannot afford to train more than one member, it is likely that he will train. Women's domestic and child-care responsibilities restrict the opportunity for training, which, like the other criteria developed for the professionalization of fish harvesters, is based on the "modern male fisher" model. This model assumes that fishers are carefree and not constrained by obligations to a fisheries-dependent family. Munk-Madsen (1998: 236) argues that the introduction of a quota system has served to formalize male control of the fishery in Norway because of women's inability to meet criteria due to female responsibilities for child care, which shape their ability to meet minimum fishing activity requirements.

In Newfoundland, there has been an increase in the number of women fishing – from 8 per cent of fish harvesters in 1981 to 18-20 per cent since 2000 (Grzetic, 2002: 115; Women in Resource Development Committee, 2003). This increase in participation rates likely reflects a number of factors, including decreased work opportunities for women with fish-plant closures and increased costs associated with operating a fishing enterprise that make family-based enterprises attractive because more money is kept in the household.

Nevertheless, it is likely that women's status as fishers will remain subordinate to men's. In other words, more women may be joining men on the boat out of economic necessity, but they are holding subordinate positions. Men, for example, hold approximately 98.4 per cent of licences and make up 92 per cent of Level II harvesters; women hold 1.6 per cent of licences and are disproportionately located at the Apprentice Level (Grzetic, 2003).

While resistance to the corporate model has been strong on the part of fishers, there has been little, if any, challenge made to the patriarchal dividend they gain by virtue of being men. In fact, there was much support for a male[5] fishery. Women who fished with fisher husbands were targets in discussions about who counts as a "genuine" fisher. Consider the following comments made by a male fisher (NGP-1):

> It could have went to a point where they went down to the committees and asked the committees on who's fishing and who's not. . . . You probably have discrimination in the sense that some fellows, he likes this one and don't like someone else . . . small communities where you got some people with their wives fishing and whatnot, it might have got core and didn't deserve, or their friend might have got core and didn't deserve it, something like that.

Note that while he is arguing that local assessments of who counts as a fisher are problematic for official use because of favouritism and discrimination, he is also implying that women fishers, particularly the wives of fisher husbands, are not to be considered genuine. Like men, women actively created a sense of difference between men's and women's paid and unpaid work. The notion of difference is not inherently problematic. However, consistent with western discourses, respondents defined differences in terms of hierarchy and inequality, perpetuating a gender inequality based on particular conceptions of masculinity and femininity and an ideology of separate spheres. Instead of becoming an opportunity for challenging gender relations, women's participation in paid fisheries work, even in the male-dominated harvesting sector, is actively interpreted in a way that reproduces the status quo.

The idea that fishing is men's work, which helps explain away women's participation, is perpetuated in a number of ways. The Newfoundland fishery has long divided work along the lines of sex, assigning men to harvesting and women to processing and household work. Until the 1980s, unemployment insurance regulations supported local custom to largely exclude women from fishing[6] (Neis, 1993: 203). In addition, the nature of sex segregation found in fish plants, positioning men in jobs associated with masculinity and literally closer to water or outside the boundaries of the factory walls and women in jobs on the production floor, distances women from fishing (Rowe, 1991: 18); and many men plant workers can claim to have fisher fathers who imparted fishing knowledge and know-how.

In reality, women do fish and have done so in Newfoundland, like in many areas of the world, historically. Despite the growing number of women entrants into the harvesting sector since the 1980s, fishing in Newfoundland is still largely men's work. This is reflected in their low rates of participation in fish harvesting. Only a small number of women actually participated in harvesting in the research area and women fishers were very much aware of being "exceptions." Consider one female fisher's reflection on her presence at a local fisheries meeting: "Yeah, I was the only woman there" (NGP-4b). When asked, most male respondents mentioned one or two independent women fishers they knew personally and evaluated their fishing performances generally very positively. For example, one male fisher (NGP-1) told me that he knew of *only one* woman who fished independently in the area, describing her as being "as good as any man was out there regards of fishing." Clearly, the standard by which women are assessed is male and the process of "exceptionalizing" women fishers allows *men and women* to continue to perceive fishing and the fishery as a male domain.

Interviews also indicated ways in which "difference" is maintained between men and women even when women are fishers. Men and women respondents pointed to the assignment of different roles and work on the boat. For example, one woman fisher told me that she left steering to her skipper husband. Work relations mirrored existing gender relations in the wider community and indicated women's subordinate role on the boat. A woman (NGP-9b) who fished with her husband described her position as being an "employee," implying, of course, that her husband was the boss: "And when I didn't quit I was fired, and the next morning I was still aboard the boat again." Despite women's presence on the boat, conservative gender relations are maintained because of women's subordinate role as crew on the boat relative to skipper husbands.

Another husband-wife fishing couple explained that fishing licences and boat ownership were registered in the husband's name and that he also dealt with buyers. Women also hold subordinate roles in this non-traditional female job because of the nature of entry, which is often dependent on co-operation from their husbands or other men. In addition, most women are relatively new to commercial fishing, so they are in subordinate positions as learners. These processes of subordination interconnect and reinforce one another: "Well, most of what I learned, most what I learned from X [my husband]. Because, but for the regard of hauling nets or anything, I didn't have a clue. . . . But I never did bother for to try to learn to steer the boat. I mean, I always had X there to do it for me" (NGP-9b, female fisher). Women's opportunities to learn are highly dependent on what men decide to impart: "I wouldn't know, X [her husband] never showed me that. He showed me the difference between a male and female lobster. And I know the difference between a male and female lump" (TEK-19b, female fisher).

Some men fishers downplayed the role their wives played on the boat in terms of physicality and strength. Typical comments included suggestions that women fishers participated only in the "easy" fisheries. One male fisher (NGP-5a) quickly made this qualification after having mentioned that his wife participated in fishing: "Yeah, some, just at this crab thing . . . crabbing is easy." Another suggested that his wife had difficulty with work involving physical strength: "She caught three [fish] on one day and she couldn't get 'em, didn't have a reel, couldn't reel. There was too much strain on to reel 'e up" (NGP-4a). The introduction of labour-easing technology, of course, makes it more difficult to exclude women on the basis of physicality. Nonetheless, there are other ways to use the physicality argument to minimize women's harvesting work. Talking to his wife, one fisher (NGP 4a) commented: "I still be at it . . . you probably finish up in probably October because it's cold, right? Too cold."

The responses by men and women do not indicate a challenge to local conservative gender relations. Rather, they help solidify the link between masculinity and fishing. Porter (1993: 135) suggests that in communities where women fish with male partners there is more overlap in the traditional sexual division of labour. She argues that the fact that women fishers work in the same physical, psychological, and economic environment as their husbands counteracts subservient relations on the boat (ibid., 145). However, this does not appear to be the case here. When we appreciate the ideological underpinnings of the ways in which local people make sense of women's paid work, that is, as being intimately connected to their household and caring work, we can begin to understand how behaviour that appears to challenge male dominance – as when women fish with male partners – actually serves to reproduce it.

Like women's work in the fish plant, their work on the boat is perceived as secondary to domestic responsibilities. The implication is that women are not considered as committed to or as invested in the fishery as men because of domestic obligations, which often translated into gaps in employment, part-time status, and a loss in seniority. This circular reasoning – where the effect (gaps in employment) is used as evidence of a lack of investment – is problematic. Nevertheless, this line of reasoning was widely accepted. One male fisher (NGP-4a) made a point of mentioning that his wife stayed ashore for such special family events as birthdays, "proving" that she was primarily committed to domestic life, which was apparently optional for him. His fisher-wife (NGP-4b) quickly responded: "One day a year, X's birthday, *that was it*. Other than that, when the boat went I went too, right?" The reality is that women are culturally assigned responsibility for domestic and caring work, which makes working for pay outside of the home more difficult for women and easier for men. This was not lost on women respondents, who pointed to the tensions between being primarily responsible for domestic and caring work and working for pay outside the home: "It was hard. I had two small kids then, right? . . . Well, a couple of summers I had a babysitter, and then between mom and his mom, we managed it" (NGP-4b, female fisher). Evidently, child care is a female problem. Despite their desires for employment or their actual training in other areas, these women were often unable to locate work outside the fishery, and they refused or were unable to leave their communities and jobs because of family commitments.

Like processing work, women's entry into the harvesting sector was explained away as a temporary or secondary household strategy: "Well, when X (her husband) gave up the draggers we needed the extra income and he had to take someone fishing so I just went" (NGP-4b, female fisher). The household strategy explanation was especially effective in distancing women in newly formed husband-wife fishing partnerships from the fishery. Women joined their fisher husbands as a strategy to deal with declining stocks and falling incomes and to cut operation costs and debt by ensuring more of the income from catches was directed to the household: "Yeah, and then I said to X (her husband), 'Well, if the two of us fishing together, it's only the two of us,' and I said, 'What we make comes into the household. You're not paying someone else.' So then they put the moratorium on cod" (NGP-9b, female fisher). Despite the fact that women's participation could be explained away, both men and women recognized the integral economic contribution women made to the household through their participation in fishing and qualification for unemployment insurance benefits: ". . . Wouldn't able to do it, right? Like I said, you know, extra income too, right? So if X (his wife) never fished, I'd have to get someone in anyway" (NGP-4a, male fisher).

Harvesting appeared to be an acceptable option when women were unable to get work elsewhere. Certainly, fishing was the only real alternative to plant work in many communities: "On times, certain years when she couldn't get work anywhere else she had no choice but to go [fishing] with me" (TEK-32, male fisher). Familial ties play an important role in women's acquisition of fishing work: "Well, there was no other job around here for her. She put in a lot of applications here and . . . there's none available. Job to get a job . . . yeah, mostly family and cousins, that's the way the fishery is around here. That's like everywhere, isn't it? I mean, all hands tries to look after their own" (NGP-5a, male fisher). Interestingly, some respondents suggested that the plant is a sexualized work environment; thus, husbands and wives fishing together was a "safer" option. Some women fishers and their male partners expressed anxiety about the possible consequences of working in such an environment. Respondents reported incidents of sexual harassment and rumours of indiscretions at the plant. In this context, women's participation in fishing was explained away as a strategy to "protect" female sexuality and marriages. One male fisher (NGP-5a) saw men's behaviour as the source of the problem at the local fish plants; yet, he also normalized men's sexual aggressiveness by suggesting that it was natural and inevitable. Viewing male sexuality in this way shaped his solutions to this problem, that is, the movement of women away from men by creating exclusive sex-segregated work environments or, alternatively, the creation of husband-wife work teams.

Women's fisheries work, particularly harvesting work, seems to have contributed to some sort of awareness about gender issues. Indeed, some men and women, fishers and plant workers, were certainly cognizant of gendered language. For example, some men fishers were aware of the pressure to use gender-neutral words such as fisher or harvester instead of "fisherman." It was clear from my interviews with men, however, that such awareness did not necessarily translate into changes in how they used language. While some respondents, primarily female, used gender-inclusive language, using terms like "fisher-*people*," most men and women respondents perpetuated the notion that women do not fish by refusing to give up the word "fisherman." In addition, women's paid participation in the harvesting and processing sectors has failed to challenge dominant gender configurations in any fundamental way.

Women respondents, like men respondents, were complicit in the perpetuation of an ideology of difference between men and women, especially in terms of work. In fact, the strength of this gender ideology comes from the active participation of women to recreate unequal relations. There are many sound reasons – ideological and material – why women voluntarily support conservative gender configurations. First, women conform to traditional norms and ideas associated with femininity, which is compromised in the factory and on the boat. Second, women are aware of a lack of opportunities for contemporary women in outport communities to experience meaningful roles as women, especially in paid employment. Instead, women carve out spaces for meaning and authority, however limited, by investing in their roles as wives and mothers.

## "YOU'VE GOT TO FIND SOMETHING FOR THEM TO DO"

Most fishers have been spending substantially less time fishing – some have withdrawn from the fishery entirely – and more time at home. Davis's research on Newfoundland (1993), Binkley's work in Nova Scotia (1995), and Gerrard's Norwegian study (1995) describe psychological and relationship-based tensions that have accompanied male fishers' increased presence at home. While displaced women respondents remained busy with domestic and caring work, men, both fishers and plant workers, complained about being bored and feeling useless, of having too much time on their hands, having "nothing to look forward to" (NGP-2), and not feeling motivated to do any non-fisheries-related tasks around the home. Some men were clearly agitated: "I mean, this is driving me off me head, you know what I mean? Being off all summer, walking around. Plus the TAGS, what you were getting on that, well, that kept you going, no doubt about it, but it's not no big money" (NGP-1, male fisher). Men plant workers mentioned similar tensions: "I don't want to be stuck around here, sitting on my ass for the rest of my life" (NGP-11, male plant worker). It was clear that because men lacked the usual structure that work provided in their lives, they sometimes felt idle or useless. Women certainly did not feel useless. After all, they still had much work to do.

A common thread running through the interviews with male respondents is the felt need to remove themselves from the home as much as possible: "Well, if the weather's good, I likes to be out" (NGP-8, male fisher). Many men combatted feelings of boredom, uselessness, and having too much time on their hands by trying to organize their lives as if they were still in full-time work or otherwise away from the house: "I spend more time away than home" (NGP-2). Men tried to frequent usual male spaces and continue to practise traditional male activities. For example, many fishers continued to occupy fishing premises where they prepared nets and boats. Often this work was unnecessary or was rooted in overly optimistic predictions about the future of the fishery and their places in it. The important point here is that men continued to meet in traditional male fisheries-related areas, but the reasons for such meetings were no longer necessarily work-related.

Reciprocal exchanges of services, including helping friends with household repairs and construction and engagement in subsistence activities, including cutting wood for fuel and growing vegetables, became increasingly important in terms of offsetting financial hardship, filling time, and providing safe spaces for men to be men. Men, individually, also carved out new spaces for themselves, including taking more active roles in their communities and local organizations. For example, in one community, local men organized the construction of a ball field. One fisher (NGP-1) described his efforts in this way: "Well, I used to be up in the woods winter time and stuff like that, but that was always, even when I was fishing, you know? . . . I took a bit more time with the union and whatnot, more work, but that was nothing really. So, no b'y, it was pretty dull I think, you know?" These efforts to keep busy and remain outside of the house demonstrate men's active attempts at maintaining continuity in their lives. The efforts also reflect the material and ideological tools available to male fishers to cope with new pressures. The "traditional male fisher" model was used to provide a recognizable order to one's day. Yet, such uses of time and space, as well as the reasons for such uses, are contrived and necessarily signify a loss. For others, this loss led to more destructive behaviours – such as drinking in sheds – that decreased their presence from the home both physically and emotionally.

Despite attempting to remove themselves from the home, men clearly spent more time in this space, doing more activities considered feminine. Some men respondents reported helping more with household work and domestic and caring tasks since 1992: "Well, when I'm home, when she's working, I'll be home and I'll cook dinner and cook supper and wash the clothes, you know? Just do a bit of the housework here and there and when she's home she does the same thing" (NGP-2, male fisher). This may reflect a link between being displaced and a pressure to do more at home, as unequal divisions of household labour may become more difficult to justify when men's work and earning power are questioned. Women's fisheries- and non-fisheries-related income and associated unemployment insurance benefits likely added to such pressure. In some cases, the wife became the breadwinner and the husband readily acknowledged his dependence on her wages, as well as the wages of other household members, in staving off welfare or out-migration. Indeed, fishers whose wives held jobs, especially in areas outside the fishery, considered themselves "lucky" (NGP-7a, male fisher). Nor did these men have any problems with women partners working for pay, especially given the current context: "It would be better if she [his wife] had a full-time job. It'd be number one then" (NGP-2, male fisher).

These disruptions of conventional gender practices hold potential for change in the household division of labour and gender relations, as well as a change in the way men and women think about gender and work. However, interview data indicate that this is not the case. Men's increased domestic participation tended to be explained away by men as a response to boredom and as a temporary arrangement. Clearly, women were still primarily responsible for this unpaid work. Men and women described this increased participation as "help," and, indeed, men's labour in the home was "supervised" in a way by women partners. In addition, men participated in these activities only when they could not find other things to do. Instead of the moratorium becoming an opportunity for more equitable gender relations, men and women interpreted changes in gender practice at the household level as temporary and by using conservative notions about what is men's and women's work. In other words, men and women were able to circumvent ideological changes despite behavioural changes. Neither men nor women interpreted the increased participation of male labour in the household as indicative of "feminization."

Retired male fishers also experienced increased time spent in the home. Like younger men who have spent more time at home since the moratorium, retired fishers reported increased participation in domestic work. Like younger men, they sought to explain away such increased participation. Because feminine work was interpreted as "helping," it was possible for retired male fishers to maintain consistent ideological understandings of masculinity despite changes in their behaviour. Doing "feminine" tasks was at times associated with the declining health of a spouse and, thus, explained away. Existing research (Keith, 1994: 143; Solomon and Szwabo, 1994: 55) suggests that while older men may take up feminine tasks, rather than this being an indication of increased role flexibility, participation in such tasks may not be integrated into their perceptions of themselves as men. This failure to integrate "feminine" work into their perceptions of themselves as men was characteristic of most male respondents, regardless of position in productive relations or age. In addition, most household tasks remained divided by gender among older couples. Older male respondents emphasized their continued ability to work hard and the persistence of traditional gender relations through an existing sexual division of labour in and around the home: "I still loves work. I still go out now before the sun come up and come in when the sun is gone down. I enjoy that. My home is while I'm eating and sleeping, you know?" (NGP-3, retired fisher, age 71). In other words, his association with his home is limited to eating meals and sleeping.

Even though most of the male fishers in this book can be described as day-trip fishers, as opposed to fishers who spend days or weeks away from families, they tended to be less visible in the home than women. If men are spending more time in traditional female spaces, what does this mean for women? Some women positively appraised the increased time spent at home by husbands. These women viewed current changes in the occupation of household space as an opportunity for husbands and wives to enjoy time together, for men to establish close family relationships, and for men to lighten women's workload in the house. However, established gender boundaries also meant that many women did not readily accept changes in the occupation of household space. Most women complained about men's presence at home:

> I find it terrible and I don't care who knows it. I mean, it's stressful. It's stressful for everybody, you know? You're just in each other's way all the time and I'm sure there's everybody find it . . . especially for somebody who's always been working. My husband's never been out of work, and I was never out of work, only for about two years, three years when the kids were born. Even then when I was home he was working, you know? . . . And I think everybody needs space, right? (NGP-17, female plant worker)

Many women found sharing the same space with husbands difficult. In a sense, these women were accustomed to being on their own. While most women respondents also worked for pay outside the home, ultimately the house was *their* space. It was a space for which they were responsible – according to the dominant separate spheres ideology – and over which they exercised some level of control in the absence of their husbands who were fishing, working opposite shifts at the local plant, or simply engaged in non-paid male activities outside the home. Women reported heightened levels of frustration since the moratorium because their husbands seemed to be always "in the way," like another "child" for whom they had to care (NGP-17, female plant worker). For women, then, the presence of husbands at home was perceived as a loss of autonomy. This is not to say that, before the moratorium, women respondents worked in unconstrained conditions inside the home. In addition to their paid jobs, women were responsible for the home and the care of children and aging and sick relatives, which created the burden of the double day. This burden meant a number of problems for women in terms of time and energy. Yet, they were autonomous in the sense that their work was self-directed.

Women respondents felt ambiguous about their domestic responsibilities, resenting being solely responsible for the home and feeling guilty when paid work interfered with their domestic performance. Consequently, the perception of autonomy comes from the self-direction; the resentment and guilt come from the culturally imposed responsibility. Viewed in this light, it is easy to understand how men's increased participation in household work has been a source of aggravation rather than relief. Women complained that the household work of male partners actually hindered their management of the home. These women had high standards for their homes and were critical of men's housework performance. Efforts to maintain "control" over domestic and caring work may also reflect a desire for continuity in their lives. Men's increased participation in domestic work, however, should not be overstated. Women remain primarily responsible for child care, domestic work, and care for the elderly. And this continuity in the lives of women does not necessarily translate into satisfaction. When asked what she did to fill her day since she was put out of work, one female plant worker explained that doing "just housework" was terribly frustrating (NGP-18).

It is important to note that female, not male, respondents raised the issue of household tension. Women, not men, described a situation where both husbands and wives were "getting on each others' nerves" (NGP-13b, female plant worker). They also explained how they had reacted to their husbands' "invasions." Many women actively found tasks outside the home for their husbands to do or assigned "safe" household tasks. For example, taking out the garbage was a safe task, as opposed to cleaning the bathroom, which required careful attention. Clearly, changes in men's work gave women another job to do – managing their husband's time. Some women encouraged their partners to continue to visit male spaces or to be engaged in traditional male activities outside the home. Alternatively, in cases where the husband was housebound for such reasons as illness or injury or was simply not interested in getting involved in activities outside the home, women removed themselves as much as possible:

> I don't be home very much. I found it difficult at first because both of us were home and we weren't used to this. Because where the shifts rotated [at the fish plant] the only time we were home together was probably Saturday and Sunday, right? That can be stressful

> when you're not used to it. But you have to adjust. . . . Some people don't, right? They just can't cope with it. But you've got to find something for them [men] to do, to get them out of the way for a while. . . . I try to make excuses, you know? "You have to do this now." (NGP-17, female plant worker)

Men's presence at home was a constant reminder for women, as it was for men, about the economic vulnerability of their households. Women's presence at home did not invoke the same visceral response, despite its potentially devastating impact on household finances. State compensation packages and early retirement options brought in less money than fishing or processing incomes, combined with unemployment insurance benefits. At the time of my interviews, many men and women had already been or were about to be cut off from financial compensation packages, intensifying this sense of vulnerability. Because of their domestic responsibilities, such cuts to the household budget often meant new or intensified work and increased anxiety and heightened burdens for women. Since the moratorium women have had to reorganize inadequate household budgets and make do with less. As members of the household are cut from compensation packages, this obviously becomes more difficult.

This situation should serve as a reminder that women's authority in and management of the household are not necessarily indicative of real power. In fact, research elsewhere (Morris, 1984: 506; 1989: 455) has shown that women's obligation for managing the household budget with insufficient income may be a source of anxiety rather than power. An inadequate budget intensifies women's domestic labour by increasing the time and effort they must put into repair and upkeep, such as mending clothing, and demands a certain inventiveness to stretch household funds and food over longer periods. One of the ways that fisheries people have been dealing with such challenges has been to "look to the past" for practical strategies such as "traditional" subsistence and reproductive activities. Such practical strategies tended to be interpreted positively in terms of the culture's unique value system. Yet recognizing women's unique household responsibilities allows us to understand that when women take up such strategies, it is not so much a choice or confirmation of maritime culture as it is an indication of economic vulnerability. Indeed, a sense of continuity and being busy may reflect burdens rather than advantage.

Women, like men, expressed tremendous concern about their standards of living, especially for their children:[7] "You haven't got the money to do what you want to do, right? . . . You can't do what you want with the kids or nothing, right?" (NGP-11, male plant worker). A male fisher (NGP-2) explained:

> Some days you're really stressed out, wondering if something happens with the mail and you don't get your [TAGS] cheque at the certain time and you got bills there to pay . . . and more worries. It's a worry, I suppose, more or less. . . . My young fellow was going to school last year in Clarenville and he was getting a training allowance but that wasn't enough to pay for everything, so, I, we had to more or less nip.

Even meeting children's basic needs has become difficult in some cases. Many women and men reported sacrifices to ensure children's needs and wants were met: "I mean, we can't afford it [socializing] the same as we did. . . . I mean now if I got twelve, fifteen bucks I got to keep it for the kids for school, you know?" (NGP-13b, female plant worker). The loss of some or all of women's and men's wages means that members of households must work harder to make do with less. Alongside this economic hardship, household members must cope with the meanings – both cultural and subjective – of being unemployed or underemployed, often with reminders from unhappy children unable to consume conspicuously.

Both men and women fisheries workers reported a sense of stigma attached to unemployment and especially financial compensation packages and a preference for opportunities to earn incomes rather than receive NCARP or TAGS. As I described earlier, women interpreted a loss of income as a loss of independence. Some men, on the other hand, reported periodically feeling inadequate and depressed for failing to provide – despite their ability to externalize blame for the fisheries crisis but not surprising given the local focus on the individual and self-sufficiency. Dependence on government support contradicted men's highly valued ideas about self-sufficiency. Male plant workers' investment in the wage is shaken by the loss of earned income. Because women's main responsibility, at least ideologically, is to manage the home, their loss of work and wages was not interpreted as a loss of femininity. Male fishers' ideas about breadwinning are complicated and connected to the type of work they do, the notion of "being satisfied," and their ability to be self-sufficient. This does not mean, however, that fishers are not aware of the pressures of consumerism and dominant ideas about masculinity defined in terms of earning cash, or that they expected their children to "sacrifice" goods and opportunities available within the wider society. Nor did fishers' reinterpretation or modification of breadwinning make financial hardship any less real.

Families could not escape these tensions that resulted in various forms of familial strain and marital breakups: "If I see somebody now that I haven't seen for a number of years, I don't dare ask about the spouse. I've after putting my foot in my mouth too often, right? And say, 'Well how is so and so?' And they say, 'Oh, didn't you know? We separated two years ago'" (NGP-17, plant worker). Some men engaged in escapist behaviour such as abusing alcohol and withdrew from families, placing additional pressure on these households. It seems unlikely that women respondents would condone such behaviour by their husbands. However, it is possible at least in some cases that women make funds available to men in order to get them out of the house. Morris's research (1984: 504; 1989: 461) on male displacement in the United Kingdom has suggested that working-class women put much effort into making money available for their husbands' personal spending, under the assumption that men's and women's needs differ and that men need to spend time and money away from home. However, she does not consider that this provision of male spending, while certainly depleting household resources, may also be a strategy women use to get men out of the house and take back female space. This point seems to be applicable in some of the cases here.

Regardless, these sorts of behaviours have led to a host of social problems at the family and community levels. Women tended to report such familial tensions much more and talked about these issues with less restraint than men. Most men I spoke with found discussing familial and social problems difficult. A few men mentioned their observations of increased problems, such as alcoholism, and suggested that men's pride worked to silence open discussions about their hard times and financial trouble: "Those guys, you don't hear them. You don't hear them talk because they're keeping it to themselves. They's out there" (NGP-1, male fisher). Another male fisher (NGP-2) stated:

> A lot of people, idle time, trying to find things to do, right? . . . I've seen it, since the moratorium come on with alcohol, right? . . . Oh yeah, they mightn't admit it, you know? If you went to a guy and asked, "Do you drink everyday?" "No, I don't know." But I've seen 'em and since the moratorium come on, seven days a week, everyday. 'Cause everybody around here got little sheds for doing their gear and one thing or another. And they're always, they go over to this guy's shed today and they probably be there for four or five days and then they go over to another guy's shed for another four or five days. Well, I even told 'em myself, I said they're going to become alcoholics. "Ah, that don't hurt. A few beer don't hurt." But see, it went from probably half a dozen between two of them per day, and now it's gone to a dozen or two dozen each per day, right?

While drinking with buddies in sheds may offer a temporary escape from the troubles of everyday life and tensions generated when men occupy female spaces, spending money on alcohol in a context of financial scarcity certainly places additional strain on the household. It also increases the work women have to do as managers of relationships and as caregivers.

### "NO PLACE FOR A WOMAN"

Coping strategies of members of fisheries-dependent households do from time to time contradict ideological constructions of gender. For example, since the moratorium men are reportedly engaging in unpaid work that has been traditionally viewed as feminine. Similarly, women have crossed gender boundaries by entering boats and engaging in breadwinning. Yet, these behavioural changes have failed to transform, at least in any fundamental way, how people think about what women and men should do and the actual gendered divisions of responsibilities. Why have women and men developed coping strategies to deal with the pressures of the moratorium that tend to perpetuate a social inequality based on previous relations and practices and particular conceptions of masculinity and femininity, reproducing the ideology of traditional gender relations? The perpetuation of the fishery as a male domain and gendered spatial geography benefit men at the local level in terms of access to fisheries resources, wealth, meanings, and identity. The reasons why women participated in such strategies that maintain the patriarchal order are complex and more contradictory. It is within the existing patriarchal order that we also find the reasons for women's active efforts to retain some sense of continuity and an explanation for their complicit reproduction of the local gender configurations.

In her research on small-scale husband-wife fishing enterprises in Norway, Munk-Madsen (1996) argues that women fishers actively support men's masculinity through voluntary submission on deck in order to confirm their femininity. This explanation is applicable to the women and men in the present study. Women fishers and processors worked in male spaces that contradicted dominant notions of femininity. Fishing and factory work are not feminine. Pollert (1981: 97) has also argued that because factory work is not feminine in the same way as other "female" jobs that are "nice, clean, respectable," women factory workers do not use work as a site to reconfirm their femininity. Of course, historical exclusion from fishing and a dominant ideology that places women in positions as consumers and housewives have contributed to this phenomenon. Despite the fact that the processing sector has been a place where the majority of workers are women, at least in the fifteen years immediately preceding the moratorium, there was a distinct feeling among men and women that the plant was either "no place for a woman" (NGP-3, retired fisher) or, at the very least, unfeminine. How could plant work be feminine? All traces of femininity had to be concealed under hairnets, overcoats, and rubber boots. Also, fish-plant work is dirty and smelly.

Many younger women processors expressed the desire to hold "clean" jobs, such as secretarial work (see also Power, 1997). Indeed, some young women held formal qualifications for such jobs but were unable to locate this kind of work in their communities or to find "clean" work that paid as well as plant work, and women could not always economically afford to stop fishing or processing. Because the boat and fish plant are clearly masculine environments and local conceptions of masculinity and femininity are based on a notion of difference, which is complementary at best, and normally hierarchical, local women and men might read women's participation in fisheries work as indicative of a loss of femininity. Women's active efforts to signify that fishing is men's work, prioritizing domestic work, and their view of fisheries work as a means to an end may be read as a strategy designed to compensate for such perceived loss of femininity. Perhaps contradictorily, however, paid fisheries work surely enabled women to gain independence by earning their own money and possibly allowed them to assert more authority over household spending. Nevertheless, because of the masculine nature of their work environments, women participated in efforts to reconfirm their attachment to the household, a space that conferred a feminine identity as well as some autonomy.

A contradiction particular to processing has to do with the fact that work is sex-segregated and that women's jobs in the factory are regarded as being effeminate for men, yet not "feminine" enough for women. Men's jobs at the fish plant, however, are distinctly "masculine." In this way, men plant workers could at least lay claim that they earned wages performing "men's work," the nature of which supposedly precluded women. For example, male plant workers suggested that heavy lifting distinguished their jobs from women's: "Well, there's some jobs like, if it's lifting, heavy work, and that kind of stuff, you can't expect a lady to do that" (NGP-10, male plant worker). However, many women provided descriptions of their jobs that hardly seemed like light work.

Somehow, women's jobs at the fish plant have become linked with their "natural abilities." Many women, in fact, used this argument to assign some value to their work, suggesting, for example, that men could not do their jobs. Pollert (1981: 97) has argued, however, that there are fundamental ideological differences between how men and women subjectively experience manual work. While factory work "can be culturally appropriated by working class men to celebrate maleness and machismo, the so-called 'light' manual work of women cannot be subjectively understood as in any way complementary to their sexual and class self-image." Instead, this work simply further devalues women.

At the same time, the ways in which women submit or defer to male authority in relation to the fishery may be an extension of their caring role – which is both self-serving and altruistic. Munk-Madsen (1996) suggests that women's efforts to reproduce the idea that fishing is a male domain can be interpreted as an extension of women's caring work. Most women respondents were very much aware of their husbands' increasingly constrained investment in the "traditional male fisher" model, as well as the "modern male fisher" model. This awareness is likely heightened at present. It is plausible that female deferral and submission are symbols of support for husbands during hard times. What Munk-Madsen (1996) fails to address is how women's behaviour and talk that perpetuate traditional gender relations are also linked to women's real-life constraints.

A history of patriarchal customs and state policies in Newfoundland has served to reproduce the idea that the fishery is a male domain. Recognizing such constraints, women were well aware of their real-life possibilities in local communities and households. Women have fewer places than men in which to invest themselves to secure meaningful and authoritative roles and identities. It is plausible, then, that women actively support the construction of a male fishery and their relegation to domestic and caring work, not only to compensate for a perceived loss of femininity or to demonstrate support for husbands, but also to cope with limited possibilities. In other words, women may voluntarily behave in ways that conform to traditional gender relations in order to appear co-operative and compliant with ideas about being a "good wife," especially when they may be perceived as stepping outside gender boundaries (for example, by fishing). In the context of unfavourable economic conditions, such conformity may be a strategy to protect some of the gains and opportunities, however limited, they have achieved in their communities and households. Local ideology about women and their work combines with their real-life possibilities to marginalize women in relation to the fishery. The idea of gender difference may be considered "beneficial" to both women and men in a society where *both* are marginalized, although in different ways and to different extents, and where both desire a sense of continuity in their lives in the context of turmoil.

## SUMMARY

In her ethnography of a rural community in Newfoundland, Davis (1993: 473) argues that ecological declines and moratoriums on a variety of species of fish and acceptance of conspicuous consumption patterns have undermined the traditional but complementary gender roles that have characterized fishing economies. The result, she argues, is the feminization of men. For Davis, evidence of this feminization includes the perception that men are more disadvantaged than women in the new economy and the increased difficulty for men to live their lives as men in traditional ways once they have become land-bound. My argument presents a rather different interpretation of the strategies and responses of Newfoundland women and men to the moratorium and ecological crisis. I concur with Davis (1993: 465-6) that the experience of being land-bound, with increased time spent at home and a loss of income and work, presents challenges for men and creates tensions in local gender relations – and certainly these tensions are played out in the home. However, women have been disproportionately hurt by neo-liberal economic trends and men gain from the patriarchal dividend. Men have tended to benefit more than women have from government adjustment programs, and they have been better able to hold on to fisheries jobs, retrain, or locate new employment.

These trends reflect women's concentration in processing, but, more importantly, they reflect the culture's imposition of the responsibility of child care and domestic work on women, which constrains employment and retraining opportunities, especially those located outside of the immediate rural locale. This is not to suggest that men have not suffered losses; clearly they have. Nor is it to suggest that all men have lost or benefited in the same ways. Differences have emerged between male plant workers and fishers, as well as among fishers, reflecting to a certain degree their relative dependence and proletarianization. Nevertheless, current professionalization and privatization efforts are also biased in their effects and support Connell's concept of a "transnational business masculinity." The state's gender-neutral language of core and non-core statuses hides how the criteria that reinforce a male fishery – albeit one that is limited and apparently more heavily integrated into capitalism – at the same time act to marginalize women. The result has been increased stress and pressure on women's time, energy, and health. Neo-liberal state policies have exacerbated the effects of job loss, at least temporarily forcing women back into the home through inadequate adjustment measures and services. Privatization and male-centred government adjustment programs have increased women's identification with the household by eliminating their other options.

# *The Winners and Losers* 7

## CONSIDERING THE LOSSES AND GAINS

The current round of restructuring in the Newfoundland fishery has enabled the increased penetration of privatization and global capitalism. The shift in values informing state direction – from an emphasis on science-based regulation to one based on classical economics – is best understood as part of a larger approach to development that has dominated Canadian fisheries policy since the 1950s. The result has been to solidify the transformation of Newfoundland fisheries from an industry once dominated by petty commodity production to one dominated by private enterprises and corporate capitalism, via the state. This corporate model is evident in the state's efforts to enclose the fisheries through boat quotas and to limit participation through the core/non-core classification system, as well as in its encouragement and support of professionalization. Who are the winners and losers in this shift?

The obvious benefactors are the corporate entities. Multinational corporations like Fishery Products International have maintained profits by restructuring, shifting harvesting effort and selling vessels to developing nations, buying fish from other suppliers, and reducing their workforce (Sinclair, 1999: 325; Ommer, 1995: 175). Corporate interests have enjoyed strong state support in terms of both policy-making and financial investment or tax breaks. Large-scale ventures have best been able to survive ecological declines because of this support, combined with their mobility and liquidity (Neis and Williams, 1997: 56; Alcock, 2000). In response to declining landings, the processing companies tied up their offshore corporate fleets and, with the assistance of the state, restructured the industry on a global scale. The result has been the closure or downsizing of the workforce in fish plants, leaving plant workers without jobs or with reduced hours and enabling companies to maintain profits. Those who have managed to retain jobs – those with the most seniority – have in some cases seen their incomes increase. The corporate model gains steam from the dismantling of the state's extensive fisheries regulatory regime and the transfer of responsibilities for management onto participants in the industry. Professionalization efforts, the core/non-core classification system, boat quotas, and enterprise allocations pave the way for the increased penetration of capital.

It is unlikely that the environment will fare well in this shift, as ecological degradation is a well-documented impact of individual quotas (see Copes, 1996). The pursuit of profit and the skewed power relations between the actors in the industry – favouring big business – have worked to the detriment of fisheries resources and local women and men dependent on these resources. Local people, in Newfoundland and around the world, have largely lost their means to meet basic needs. Newfoundland regions with a recent history of crab and shrimp harvesting and processing are coping better than those heavily dependent on the groundfishery. Communities with access to shellfish harvesting and processing are pitted against communities lacking such access in a struggle to survive. Communities fracture internally as workers fight for jobs and access to scarce resources. While incomes of fisheries workers have increased on average, the number of people employed in the fishing industry has declined. Households with non-fisheries incomes have fared better than those entirely dependent on the fishery.

Perhaps it is women who have lost the most. State initiatives have advantaged men over women through an unequal distribution of wealth and resources, both social and material. Women faced a lack of available work, unsuitable retraining opportunities, and lower replacement incomes for shorter durations; yet, men *felt* more disadvantaged than women. A crisis of fish has not translated into a crisis of masculinity among men fishers. This would require a breakdown in patriarchy, which has not happened. Individual men may feel powerless; yet, practice has been configured in ways that confer space for men to exercise power. Women have been disproportionately hurt by neo-liberal economic trends and men gain from the patriarchal dividend. Social programs and the introduction of intensive forms of capitalist accumulation have reflected and reinforced gendered divisions of labour and notions of women's and men's proper places in what is defined as an all-male extraction industry. Men have tended to benefit more from government adjustment programs, and have been better able to hold on to fisheries jobs, retrain, or locate new employment than women. These trends reflect the gendered divisions of labour in the fishery, where women are concentrated in processing and men in harvesting. Processing has been more negatively affected than harvesting by reduced quotas, increased automation, and the production of a less labour-intensive product in crab and shellfish processing.

Women's disadvantageous position, however, cannot be entirely reduced to their concentration in processing. Rather, it reflects the culture's imposition of the responsibility of child care and domestic work on women, which constrains employment and training opportunities, especially those located outside of the immediate rural locale. Assumptions about women's responsibilities inform the state's fisheries policies. The gender-neutral language of core and non-core statuses hides how the criteria marginalize women and reinforce a male fishery, albeit one that is limited and apparently more heavily integrated into capitalism. As neo-liberal policies compel households to do more subsistence and unpaid work to meet the needs of their members, not all members equally share this responsibility or have equal access to resources. Such policies have resulted in gendered stressors in fishing communities and households, along with increased pressure on women's time and health and an intensification of their work. Neo-liberal state policies exacerbated the effects of job loss, at least temporarily forcing women back into the home through inadequate adjustment measures and services. Male and other household members benefit from women's intensified domestic and caring labour and women become more heavily dependent on male wages. Privatization and male-centred government adjustment programs increased women's identification with the household by eliminating their other options, in turn, freeing men's time to invest in fishing – ideologically and in practice – and supporting a "transnational business masculinity."

That is not to suggest that men have not suffered losses; they have. Nor is it to suggest that all men have lost or benefited in the same ways. Differences have emerged between male plant workers and fishers, as well as among fishers, reflecting the different positions in production. These differences can be explained, in part, by the relative dependence and proletarianization of the various groups. Current professionalization and downsizing efforts support male exclusivity and encourage the penetration and intensification of corporate imperatives within the fishery, weakening the position of the smallest capital interests and forcing a technological race to remain economically viable in the face of resource decline. The shift to harvesting other species has been particularly detrimental for the small-boat sector, especially those without crab permits and, therefore, without access to fisheries wealth. Steady declines in the total number of vessels, especially those under thirty-five feet, indicate this shift. There are few, if any, real alternatives to fisheries work in rural outports. Locating alternative work in urban areas is not a likely option given many fishers' low educational levels and their commitment to rurality. Fishers, then, must choose between "going bigger" or getting out, and those with access to financial resources will benefit most. Obstacles to such investment in technological advancement are not only financial but also relate to ideas – particularly perceptions about what a "real" fisher is and does – as well as to changes in work processes and conditions. Fishers' possibilities and decisions may also affect sons' decisions regarding entry, the composition of the next generation of fishers in Newfoundland, and the conditions under which they work.

## COPING, RESISTANCE, AND CHANGE

Faludi (1999: 86) argues that men who perform work that is embedded in a social or historical context, and that therefore has greater meaning and significance in their lives, are better able to negotiate displacement and unemployment. Small-boat fishers were certainly able to draw on their particular positions as primary producers, a masculine maritime heritage, and their history of political marginalization and local struggles to develop a framework in which to understand change, negotiate tensions, and cope. This framework has allowed men to assign blame for the current crisis outside of themselves and to realize that few (if any) other work opportunities exist that could offer the same sorts of meanings and rewards as fishing. This framework, this discourse, makes fishing worth clinging to. Kaufman (1994: 145) has suggested that "[e]ach subgroup, based on race, class, sexual orientation, or whatever, defines manhood in ways that conform to the economic and social possibilities of that group." Available material resources and discourses shape people's daily lives, their identity formation, and how they make sense of the world, and these ways of seeing oneself in the world can either inhibit or enable social action.

Women and men make decisions about how to cope with unemployment and economic vulnerability not as isolated individuals but in social contexts and as part of social groups, and possibilities in both the formal and informal sectors of the economy are enabled and constrained by accepted ideas and practices. Equally, the decisions of women and men are shaped by their own class and gender positions, as well as those of other household members. The household is central to understanding why people behave in certain ways and not others. In other words, individual work histories, the work patterns and strategies of household members, and potential conflicts between members influence class identification, action, and inaction.

Small-boat male fishers have largely resisted state-directed efforts to downsize the fishery, engaging in a number of coping strategies that reflect culturally specific gender and class configurations of practice and ideas. In their attempts to maintain some sort of continuity in their lives and to get by, small-boat fishers drew on lessons learned from "the collective past." In fact, not resisting was largely unthinkable because fishing is "in the blood." "Looking to the past" and drawing on the "traditional male fisher" model provide tools – material and ideational – with which to respond and develop a framework to make sense of restructuring and associated changes. This framework, combined with fishers' position in productive relations, helped to organize coping strategies, including the adaptation of usual fishing strategies, the engagement in illegal or informal economic activity, the intensification of subsistence work, and the rejection of opportunities to train for non-fisheries related work. Local discourses that privilege fishers' knowledge made possible the construction of oppositional explanations and culturally meaningful male identities and constrained participation in formal training. These responses reflect structured opportunities available to *men* and to *fishers*, as well as constraints; they also reflect male agency and the importance of the local in mediating the effects of restructuring.

Inshore male fishers had access to alternative discourses and were thus able to connect their problems to society. This connection is linked to political social action. Small-boat fishers have tended to earn low incomes, periodically necessitating other sources of employment, and depend heavily on local processors to buy their product. It is in this semi-proletarianized position that there is potential for collective action and divisiveness, opposition and acceptance of capitalist integration. While fishers experience many of the vulnerabilities of the working class and are often preoccupied with wages, they do not always identify themselves as working class or proletariat. In fact, rather than identify with plant workers, fishers saw their work as fundamentally different. Further, still, the "traditional male fisher" model is subversive in its conception and evaluation of time, men's work, and breadwinning, yet reactionary in its assumption about its maleness. This assumption serves to separate fishers from potential allies in women. Potential for resistance is also curbed as age is used by fishers to define exclusive group membership. Despite the widespread similarity, the coping strategies described tended to be individualistic and family-oriented. Cultural constructs including the notions of "self-sufficiency" and "being satisfied" support the neo-liberal agenda. At the same time, inshore fishers' marginal position in relation to capital permits the development of a pointed critique of government-led fisheries management and science.

The contradictory class position of inshore fishers appears to be enabling, constraining, and mediating changes in local configurations of practice and ways of thinking. The "modern male fisher" model is challenging the local hegemonic version of masculinity – the "traditional male fisher." This challenge is particularly significant because it comes from both outside (the state) and inside the local culture. The success of this challenge to the more subversive "traditional" model is undoubtedly linked to fishers' material conditions. Declining stocks and incomes will boost support for individual quotas to secure access to fisheries wealth and for individual transferable quotas to allow fishers to sell and exit an industry in trouble. These resource exigencies and policy determinations, which are beyond the direct control of the people of this study, will likely undermine commitment to an open fishery based on egalitarianism, and they will serve as impediments to collective action. At the same time, these factors will further fragment and fracture communities that have been under stress for many years.

Fishers were divided regarding the value of science and the extent to which inshore fishers themselves are to blame; they were also divided along the line of age. Much of this convoluted debate is dedicated to portraying oneself in the most favourable light possible. When fishers accuse *other* fishers of being partially responsible for the fisheries crisis or for having it too easy, they tend to do so by using themselves as the standard by which others should be measured. Similarly, debates over the usefulness of science in fisheries management may reflect a desire to present oneself in a particular way, perhaps more "enlightened," more "modern," challenging the "traditional male fisher" model, even if only partially. These individualistic and divisive responses reflect fishers' concerns about their future involvement in the fishery and the often unclear direction of the state. Individual male fishers have responded with support for and resistance to this direction in their efforts to claim a place in the future downsized fishery. Drawing on two competing local discourses – "looking to the past" and "looking forward" – and accepting the "too many fishers, too few fish" argument, fishers argued for further enclosure, increased state management, or both. Either way, fishers developed criteria for inclusion and exclusion in a future fishery that tended to reflect their positions. These discourses and models are so important because they are about both access to and the distribution and control of ideas and resources.

## THE LOCAL AND THE GLOBAL

The New Right political and neo-liberal economic agendas have become the dominant approach for addressing development and crises in Canada. This model for fisheries – marked by market forces, decentralization, privatization, and government withdrawal – is a global trend (Skaptadottir and Proppe, 2000), and the state uses this trend to legitimate its direction. The implication is that we must not fall behind and it is better to jump on board the bandwagon now before it is too late. For example, *The Report of the Independent Panel on Access Criteria* (Department of Fisheries and Oceans, 2002: 31-4) selectively lists decisions regarding access issues in other countries, including New Zealand, Iceland, and Australia, all of which have moved towards ITQs or versions thereof, and cases from the European Union and the United States that demonstrate a closing of the commons. It is then noted: "Granting access, essentially free of charge, as is done in Atlantic Canadian fisheries experiencing an increase in resource abundance and/or landed value, is highly unusual in other jurisdictions and sectors. In other jurisdictions, fishers wishing to gain access to a fishery have to buy their way in" (Department of Fisheries and Oceans, 2002: 37).

While the precise consequences of neo-liberal policies on local people are influenced and constrained by the specific historical, political, and cultural contexts of nations and regions, the research suggests that increased difficulties meeting basic needs, declining health, and accentuated inequalities appear to be trends in economies where fisheries are undergoing enclosure. Such problems are exacerbated for women because neo-liberal models for the direction of the fishery simultaneously ignore, reproduce, and assume the existence of gendered divisions of labour and inequalities. It is certain that we will not find the prerequisites for sustainable communities and fair economic organization in approaches that rely on gender or other inequalities.

# Appendix

## *Methodology: Some Concerns*

As a Newfoundlander, I am deeply concerned about the future of local people and their communities in light of chronic unemployment. As a feminist, I am particularly concerned with how the new burdens associated with the moratorium will impact on the gendered lives of women and men. These concerns and connections led me to ask questions about the nature of women's and men's experiences of and responses to the most recent fisheries crisis. In particular, I was concerned with how the work experiences of male inshore fishers affected their responses to tensions created by the threat of displacement and restricted access, as well as how the coping strategies of men were linked to local gender configurations of practice and ideologies. This brings me to my first methodological concern regarding this work, namely, the concentration on one occupational group. Any attempt at accounting for and analyzing a particular occupational group of men, like male inshore fishers in Newfoundland, runs the risk of idealization or stereotyping by focusing on representative characteristics and prioritizing economic position over other social positions in which men find themselves. I have tried to manage this problem by simultaneously presenting commonalities and differences within this group.

A second methodological concern deals with the significance of the timing of this study. I conducted interviews between five and six years after the moratorium on northern cod had been declared, and the TEK interviews took place between three and four years after 1992. Responses surely reflect this timing. For example, most respondents were still receiving state compensation or had just been cut off from such financial packages at the time of the interviews and felt that the worst was yet to come. Financial compensation packages may have buffered some of the more negative economic effects of the fishing bans on cod and other groundfish, masking or postponing potential changes in local gender constructions and relations, as well as the full psychological impact.

Other methodological issues include the reliability of my sample, the information that respondents provided, and my interpretation of the data given the politically charged context (Judd et al., 1991: 259). I compiled my sample from a personal list of contact names, which grew through recommendations from respondents. This approach is problematic because it does not produce a random sample and may lead to the omission of divergent perspectives held by those who failed or refused to be questioned. However, I encountered little resistance to questioning. No doubt accessing respondents and generating trust were linked to my status as an insider, having grown up in a rural Newfoundland community. Yet, this status has presented unique quandaries. For example, my interpretations of the data may be coloured by my love-hate relationship with Newfoundland. It is easy to romanticize life in Newfoundland, especially in terms of the opportunities available to explore and connect with nature, something that the relatively new tourist industry has vigorously marketed. Simultaneously, I have long felt that rural Newfoundland can be a very socially oppressive environment for women, given the accepted gendered divisions of labour and ideologies. Moreover, the possible political ramifications of reflecting negatively on a relatively powerless group of people engaged in a constant struggle for financial compensation and other resources from the state may also hinder, unintentionally or otherwise, my attempt at honest reflection. This politically charged context may have encouraged interviewees to be careful in how they responded to my questions and might explain why some respondents refused to be taped, despite agreements of confidentiality. Even in the absence of such a highly charged climate, tape recorders are often regarded as intrusive (ibid.). Some fisheries workers were suspicious about my intentions and feared disclosing sensitive information that could potentially be used against them or used to legitimize cuts in federal income replacement programs. In addition, people from this region of Newfoundland are no strangers to the research process and some respondents expressed discouragement with the lack of improvements made to their own lives and communities despite all of the ecological and social research done in the area.

I would be remiss if I ignored the potential methodological problems presented by a woman studying masculinities. According to Walker (1994: 41-2), men's responses may be both positively and negatively influenced by an interviewer's sex. Female interviewers may be more effective than male interviewers at eliciting honest answers from male respondents regarding distinctions between real and ideal masculine practices. On the other hand, men may be more likely to choose their words carefully in ways that exaggerate their acceptance of feminist ideas when interviewed by women rather than men. In addition, women interviewers are often met with comments or gestures acknowledging their sex. For example, it was not uncommon for male fisheries workers in this study to edit their responses, using qualifying statements such as: "I'm not a chauvinist, but . . ." (NGP-7a, male fisher). Equally, however, respondents commented on my educational status. For example, when asked his views on formal science and fisheries scientists, one male plant worker (NGP-11) responded by saying: "I don't think too much of the scientists. That's not where *you're* going to work, is it?" Some men were inclined to be patronizing, perhaps unintentionally, and expected me to know very little about their masculine world of work. A positive side to this male assumption was that it gave me "permission" to ask for further clarification and details when necessary. It is likely that men perceived me to be non-threatening because of my sex, student status (at the time of the interviews), and relative younger age. My observational work in "traditional male spaces," such as the Port Union wharf or men's fishing premises, also elicited gendered responses. Indeed, I was the object of curiosity – and an occasional catcall – in these places. However, the curiosity I generated may be attributed less to my sex and more to the fact that I was a stranger whose intentions were unknown. On the other hand, talk in outport Newfoundland travels fast and it is unlikely that much time had passed before most of the locals were aware that there was a researcher in the area, which may in itself have generated curiosity and distrust. My gender-related methodological problems, however, were not limited to interviewing or interacting with men. In fact, I found women more difficult to interview than men. This reflects my topic of research, the fishery, and women's marginal status in fishing communities. Women tended to think that questions about the fishery should be directed to their menfolk. Such behaviour in the interviewing process, however, provided evidence of larger ideologies at work in everyday life.

This book is not the definitive story of men inshore fishers in Newfoundland. Rather, it is one interpretation of the data – one that I hope inspires further discussion and research.

# Notes

*Chapter 1*

1. The island of Newfoundland and the mainland region of Labrador make up the province of Newfoundland and Labrador. Until 2002, the province's official name was Newfoundland. This study was conducted on the island portion of the province. However, all statistics describe provincial figures. The cod moratorium also affected many coastal communities in Labrador.
2. See Appendix, Methodology: Some Concerns, for a review of some methodological problems and limitations.

*Chapter 2*

1. The European migratory fishery dates back to the fifteenth century (Hutchings and Myers, 1995: 41). Obstacles to permanent settlement by Europeans included resistance from merchants and the British government and the lack of women able or willing to settle the island (McGrath et al., 1995: 22-6; Porter, 1993: 3, 42).
2. Neis (1993: 190) reminds us that female participation in the onshore production of dried salted fish varied from region to region. In addition, processing techniques and quality varied across households and communities, as well as with varieties of fish caught (Porter, 1993: 47). See Ferguson (1996) for a detailed description of salting and drying techniques in Newfoundland.
3. These markets were located mostly in Europe, the Caribbean, and Brazil (Sinclair, 1988b: 158).
4. Withdrawal from the salt fishery occurred before Confederation in some areas of the province (see Stiles, 1972: 37).
5. "Outport" is a term used to describe remote fishing villages in Newfoundland.

6. See Power (1997, 2000) for discussions of unsustainable company practices, including the use of wasteful processing machines and off-shore harvesting methods.
7. As Neis (1992: 157) documents, inshore fishers in Newfoundland have voiced concerns about the health of various fish stocks since the 1980s. This is not a new phenomenon. Recent historical evidence suggests that fishers and others have made similar articulations since the early 1800s, especially with regard to the introduction of new, more efficient technologies (Ommer, 1998: 14).
8 NAFO has developed geographical zones for the purpose of managing northern cod stocks, including Division 2J, which is located in the waters off southern Labrador, 3K, on the northeast Newfoundland Shelf, and 3L, which refers to the northern Grand Banks. Collectively they are known as 2J3KL (Hutchings and Myers, 1995: 41).
9. The federal government has jurisdiction over harvesting; the provincial government licenses processing plants.
10. Originally TAGS was more comprehensive, including efforts to professionalize fishers. It also expected clients to be "active," a criterion that was later dropped (see Human Resources Development Canada, 1998: ch. 2).
11. I was a researcher with this project between 1995 and 1996 and am grateful to Barbara Neis and Lawrence Felt for allowing me access to its data.
12. Some of the TEK interview transcripts did not supply complete information regarding marital status, number of children, spouse's occupation, education, and age.
13. The low number of women fisher respondents reflects the low number of women who fish for a living in the area, as well as throughout the province.

*Chapter 3*

1. Much research on Newfoundland has described local resistance to change and outside norms. Nemec (1972), for example, reports a tendency for inshore fishers to reject modifications to fishing, technological or otherwise, that endanger culturally valued relationships or organization.
2. A blue note is a "receipt for fish sold to a merchant, used as credit for goods and provisions to be purchased" (*Dictionary of Newfoundland English*, 1999).
3. Unemployment insurance benefits based on fishing income tend to be perceived more favourably than UI benefits based on income from make-work projects or from state welfare payments: "We've always lived. We always survived. And it's one thing I'll say here in this community, there's never a make-work project come here since I've known

it to help anyone get their stamps. We've always managed. Lots of times, they were low. But you survived" (TEK-33, male fisher).

4. A felt sense of unity, however, does not imply the absence of interpersonal problems at the community level or the extension of endearment to all members of a community. See Szwed (1966), Felt (1987), Wadel (1973).
5. In 1988, 84 per cent of the 17,786 individuals in Newfoundland and Labrador who reported earnings from fishing had total incomes of less than $20,000. About one-third had total incomes of less than $10,000. Note that these total incomes include fishing income, non-fishing income, unemployment insurance benefits, and other taxable income (Economic Recovery Commission, 1993: 3). See also ibid., ch. 2, "Economic and Population Trends Post-Moratoria."
6. The notion of egalitarianism in Newfoundland villages has been explored in a variety of ways by different authors. For example, Davis (1988) has examined how locals assessed, negatively or positively, women's efforts to look after their families in terms of this ethic. Nemec (1972) and Stiles (1972) have demonstrated how this community-based ethic mediates relationships among crew members.
7. These decent wages reflect union efforts and are not representative of wages in all fish plants in the province.

## *Chapter 4*

1. Brian Tobin was the Premier of the province of Newfoundland when I conducted my interviews and has also acted as federal Minister of Fisheries.
2. These fisheries are considered "wasteful" because they are prosecuted primarily for roe.
3. A ghost net is a "fish-net lost through storms or neglect," which continues to catch fish unchecked until the materials decompose (*Dictionary of Newfoundland English,* 1999).
4. Vic Young was the chief executive of Fishery Products International at the time of this interview.
5. People in the research area often referred to the plant located in Port Union as the Catalina plant because of its location on its border.
6. Fishers did not exclusively use *fisheries* scientists to construct a sense of differentness. Indeed, scientists more generally, including academic researchers, were targets. A 46-year-old fisher (TEK-27) illustrated this point when a TEK researcher asked for clarification on a comment he had made by responding, "I don't know how to explain it in *your language*" (emphasis my own).
7. Of course, women and plant workers also have knowledge about the fishery. Elsewhere I have argued that understanding women process-

ing workers' experiences and the inherent contradictions and tensions of processing fish for sale and also of procuring and preparing it for home consumption can help establish the necessary conditions for sustainable fisheries (Power, 1997, 2000). They have knowledge about changes in the species composition of catches, location of catches, and the size of fish and how these changes interact with the organization of work in processing, plus they are familiar with consumer markets that influence their working lives and food options. In addition, the sexual division of labour within fishing households affects knowledge relevant to understanding fisheries and fish ecologies. Men fishers know more about fish ecology and fishing practices, but their wives often know more clearly when changes in fishing vessels and gear occurred and maintain the financial records where such information is recorded. This information is extremely important for collecting, verifying, and interpreting information from fishers about fish ecology and abundance trends.

8. A popular focus of research in the early literature on men fishers was crew composition and recruitment, patrilineal inheritance customs, and patrilocal residence patterns, especially with regard to the impact of modernization and resettlement schemes on these patterns (Andersen and Wadel, 1972a, 1972b; Nemec, 1972). The exact shape of crews depended on whether or not inherited fishing gear was divisible or not (Andersen and Wadel, 1972b: 147-9). Fishing gear that could not be divided up encouraged crews composed of brothers; divisible gear encouraged temporary sibling-based crews. When economically viable, brothers divided gear and formed separate enterprises.

## *Chapter 5*

1. This term refers generally to people from St. John's, the capital city of Newfoundland and Labrador.
2. Although fishers and processing workers expressed skepticism regarding the usefulness of fisheries science and ambivalence towards scientists, they also tended to be very supportive of local youth pursuing higher education. My positive research experience certainly reflected this support. While fishers invest in and value a particular type of "fishing education," they also respect and appreciate the more formalized educational choices of their children and younger generations. These sentiments, of course, exist in contradiction and sometimes in tension with each other.
3. A number of studies have shown how local cultural values mediate the adoption and use of particular strategies and technologies in the post-Confederation period. For example, Stiles (1972) found that an egalitarian ethic mediated fishers' use of new radio communication technology.
4. This sort of criticism parallels the pre-moratorium criticism that some fishers simply "fished for stamps." The term "stamps" refers to a

"printed slip pasted on government form administered by an employer, recording amount of unemployment benefits due to employee, esp a fisherman" (*Dictionary of Newfoundland English*, 1999).

5. Men plant workers also resisted retirement. A male plant worker (NGP-12a) told me a story about a 75-year-old man who refused to retire from the plant: "He thinks he's going to die if he gives up work."
6. At the time of this study, Earle McCurdy held the position of FFAW president and Reg Anstey was FFAW secretary-treasurer.
7. Prior to the decline in the fishery, men and women plant workers did exercise resistance at work, for example, through the manipulation of the bonus system (Power, 1997). Such instances of worker resistance were hampered by job scarcity and changes in the raw material, which made it difficult to meet performance requirements. For those still working, it appears that resistance was replaced with compliance.

*Chapter 6*

* Modified versions of parts of this chapter appear in the following forthcoming works: 2005, with Donna Harrison, "Lessons from India: Applying a 'Third World' Framework to Examine the Impacts of Fisheries Crisis on Women in Newfoundland Villages," *Advances in Gender Research: Gender Realities: Local and Global*, Vol. 9; 2005, with Donna Harrison, "The Power of Gender Ideology in the Face of Resource Decline in Newfoundland, Canada" in *Changing Tides: Gender, Fisheries, and Globalisation*, edited by Neis et al., Fernwood and Zed Books; and 2005, "The Newfoundland Fishery: 'No Place for a Woman'?" in *Gendered Intersections: A Collection of Readings for Women's and Gender Studies*, edited by L. Biggs and P. Downe, Fernwood Press.

1. This work, however, generated fears about the potentially negative impact fisheries work and time at sea would have on women's femininity, sexuality, morality, and health (McGrath et al., 1995: 26).
2. Interviews with fishers illustrated this point: "Probably if the wife was here she'd get it [fishing receipts]. She knows where it is at. I don't know" (TEK-3, male fisher). During my interviews with married couples, men often gave their background information sheets to their wives to complete on their behalf and referred to them for dates and other information.
3. This woman clearly expected me to understand the pressures placed on women to keep up with both paid work outside the home and unpaid domestic and caring work.
4. Research on women and processing work suggests that the increase in the number of women gaining employment in plants during the 1970s and early 1980s was accompanied by an intensification of sex segregation (Rowe, 1991: 15).

5. While most men did not blame women for ecological degradation, we cannot rule out the possibility that some men and policy-makers linked women's increased participation in the harvesting sector to stock declines. For example, one retired male fisher (TEK-34) attributed the entry of women in the harvesting sector as contributing to the problem of "too many fishing for too few fish" and taking away work from men: "Since '65 when we came here, how many women on Newfoundland island got a fishing licence? . . . Should never have been." Men I spoke with may have consciously avoided making such claims in order to dodge the label "sexist." In addition, to suggest that women could somehow be partially responsible for the current crisis is problematic for fishers who also relegate women to the domestic sphere, a strategy that marginalizes women's involvement in the fishery.
6. Between 1957 and 1958 unemployment insurance benefits, designed with input from federal fisheries officers, were extended to include fishing work performed by men. Until 1981 women who fished with husbands or processed fish as part of the household-based fishery did not qualify. Instead, male family members were credited with the fisheries work that women had done (Neis, 1993: 203).
7. I should note that some men respondents denied the negative economic impact of moratoriums on the community and cited physical evidence, including the construction of new houses, to refute such suggestions. When I asked about his and others' material well-being, one male plant worker (NGP-12a) was quick to point out the window to show me just how well people were doing: "Yes, this place is all built up here now." He later conceded that there may indeed be growing poverty in the area, but that such problems are concealed.

# *Bibliography*


Alcock, F. 2000. "Scale Conflict and Sectoral Crisis: A Comparative Analysis of Canadian and Norwegian Cod Industries," paper presented at the XVIIIth World Congress of the International Political Science Association, Quebec City, 4 Aug.

Andersen, R. 1972. "Hunt and Deceive: Information Management in Newfoundland Deep-Sea Trawler Fishing," in R. Andersen and C. Wadel, eds., *North Atlantic Fishermen: Anthropological Essays on Modern Fishing*. St. John's: ISER Books.

——— and C. Wadel. 1972a. "Introduction," in Andersen and Wadel, eds., *North Atlantic Fishermen.*

——— and ———. 1972b. "Comparative Problems in Fishing Adaptations," in Andersen and Wadel, eds., *North Atlantic Fishermen.*

Antler, E., and J. Faris. 1979. "Adaptation to Changes in Technology and Government Policy: A Newfoundland Example," in R. Andersen, ed., *North Atlantic Maritime Cultures*. The Hague: Mouton.

Armstrong, P. 1996. "The Feminization of the Labour Force: Harmonizing Down in a Global Economy," in I. Bakkar, ed., *Rethinking Restructuring: Gender and Change in Canada.* Toronto: University of Toronto Press.

Bakkar, I. 1996. "Introduction: The Gendered Foundations of Restructuring in Canada," in I. Bakkar, ed., *Rethinking Restructuring: Gender and Change in Canada.* Toronto: University of Toronto Press.

Barnes, B. 1985. *About Science.* New York: Basil Blackwell.

Binkley, M. 1995. "Lost Moorings: Offshore Fishing Families Coping with the Fisheries Crisis," *Dalhousie Law Journal* 18, 1: 84-95.

Bird, S.R. 1996. "Welcome to the Men's Club: Homosociality and the Maintenance of Hegemonic Masculinity," *Gender & Society* 10, 2: 120-32.

Bly, R. 1990. *Iron John: A Book about Men.* Reading, Mass.: Addison-Wesley.

Brittan, A. 1989. *Masculinity and Power.* Oxford: Basil Blackwell.

Brod, H. 1994. "Some Thoughts on Some Histories of Some Masculinities: Jews and Other Others," in Brod and Kaufman (1994).

——— and M. Kaufman, eds. 1994. *Theorizing Masculinities: Research on Men and Masculinities.* Thousand Oaks, Calif.: Sage.

Brodie, J. 1996. "Restructuring and the New Citizenship," in I. Bakkar, ed., *Rethinking Restructuring: Gender and Change in Canada.* Toronto: University of Toronto Press.

Cadigan, S. 1995. "Whipping Them into Shape: State Refinement of Patriarchy among Conception Bay Fishing Families, 1787-1825," in C. McGrath, B. Neis, and M. Porter, eds., *Their Lives and Times: Women in Newfoundland and Labrador, A Collage.* St. John's: Killick Press.

Canadian Council of Professional Fish Harvesters. 2001. "Preliminary Response of the Canadian Council of Professional Fish Harvesters to The Atlantic Fisheries Policy Review." At: www.ccpfh-ccpp.org/eng/pubaccueil.html

Christie, A. 2001. "Men as Social Workers in the UK: Professional Locations and Gendering Global Discourses of Welfare," in B. Pease and K. Pringle, eds., *A Man's World? Changing Men's Practices in a Globalized World.* London: Zed Books.

Clatterbaugh, K. 1996. "Are Men Oppressed?" in L. May, R. Strikwerda, and P. Hopkins, eds., *Rethinking Masculinity: Philosophical Explorations in Light of Feminism,* 2nd ed. Lanham, Md.: Rowman & Littlefield.

———. 1997. *Contemporary Perspective on Masculinity: Men, Women, and Politics in Modern Society,* 2nd ed. Oxford: Westview Press.

Cleaver, F. 2002. "Men and Masculinities: New Directions in Gender and Development," in F. Cleaver, ed., *Masculinities Matter! Men, Gender and Development.* London: Zed Books.

Cohen, M.G. 1997. "What Women Should Know about Economic Fundamentalism," *Atlantis* 21, 2 (Spring/Summer): 4-15.

Collinson, D.L. 1992. *Managing the Shopfloor: Subjectivity, Masculinity and Workplace Culture.* Berlin: Walter de Gruyter.

——— and J. Hearn, eds. 1996. *Men as Managers, Managers as Men: Critical Perspectives on Men, Masculinities and Managements.* London: Sage.

Coltrane, S. 1994. "Theorizing Masculinities in Contemporary Social Science," in Brod and Kaufman (1994).

Connell, R.W. 1995. *Masculinities.* Oxford: Polity Press.

———. 1998. "Masculinities and Globalization," *Men and Masculinities* 1, 1: 3-23.

Connelly, M.P, T.M. Li, M. MacDonald, and J.L. Parpart. 2000. "Feminism and Development: Theoretical Perspectives," in J.L. Parpart, M.P. Connelly, and V.E. Barriteau, eds., *Theoretical Perspectives on Gender and Development.* Ottawa: International Development Research Centre.

——— and M. MacDonald. 1996. "The Labour Market, the State, and the Reorganization of Work: Policy Impacts," in I. Bakkar, ed., *Rethinking Restructuring: Gender and Change in Canada*. Toronto: University of Toronto Press.

Copes, P. 1996. "Adverse Impacts of Individual Quota Systems on Conservation and Fish Harvest Productivity." Discussion Paper 96-1. Burnaby, B.C.: Institute of Fisheries Analysis, Simon Fraser University.

Cornwall, A. 1997. "Men, Masculinity and 'Gender in Development'," *Gender and Development* 5, 2 (June): 8-12.

Davis, D.L. 1983. "Woman the Worrier: Confronting Feminist and Biomedical Archetypes of Stress," *Women's Studies* 10: 135-46.

———. 1988. "'Shore Skippers' and 'Grass Widows': Active and Passive Women's Roles in a Newfoundland Fishery," in Nadel-Klein and Davis (1988).

———. 1993. "When Men Become 'Women': Gender Antagonism and the Changing Sexual Geography of Work in Newfoundland," *Sex Roles* 29, 7/8: 457-75.

Department of Fisheries and Aquaculture. 1998. T*he 1997 Newfoundland and Labrador Seafood Industry – Year in Review*. At: http://www.gov.nf.ca/Fishaq/publications/yir_1997.stm

———. 2001. *2000 Seafood Industry Year in Review*. St. John's: Government of Newfoundland and Labrador.

———. 2003a. Web site. At: www.gov.nf.ca/Fishaq

———. 2003b. *2002 Seafood Industry in Review*. St. John's: Government of Newfoundland and Labrador.

Department of Fisheries and Oceans. 1996. Commercial Fisheries Licensing Policy for Eastern Canada, Ottawa: Communications Directorate. At: http://www.dfo-mpo.gc.ca/communic/lic_pol/index_e.htm

———. 1997. *Newfoundland and Labrador Inshore Sentinel Survey: Fishers and Scientists Working Together*. Ottawa: Communications Directorate.

———. 2001. Fisheries Management Policies on Canada's Atlantic Coast. At: http://www.dfo-mpo.gc.ca/afpr-rppa/FINAL%20-%20ENGLISH%20-%20APRIL%202002.pdf

———. 2002. *Report of the Independent Panel on Access Criteria for the Atlantic Coast Commercial Fishery*. Ottawa: Government of Canada.

———. 2003a. Licensing Statistics: Species Information. At: www.dfo-mpo.gc.ca/communic/statistics/LICENSES/licinfoe.htm

———. 2003b. Licensing Statistics: Vessel Information. At: www.dfo-mpo.gc.ca/communic/statistics/LICENSES/vessel_e.htm

———. 2003c. Licensing Statistics: Species Information. At: www.dfo-mpo.gc.ca/communic/statistics/LICENSES/fisher_e.htm

———. 2003d. Canadian Landings Information. At: www.dfo-mpo.gc.ca /communic/statistics/landings/land_e.htm

Dictionary of Newfoundland English, 1999. At: www.heritage.nf.ca/dictionary

Economic Recovery Commission. 1993. *An Overview of Incomes of Workers in the Newfoundland and Labrador Fishing Industry.* St. John's: Economic Recovery Commission, Newfoundland and Labrador.

Economics and Statistics Branch. 2003. T*he Economy: The New Newfoundland and Labrador.* St. John's: Department of Finance, Government of Newfoundland and Labrador.

Edley, N., and M. Wetherell. 1995. *Men in Perspective: Practice, Power and Identity.* London: Prentice-Hall/Harvester Wheatsheaf.

Faludi, S. 1999. *Stiffed: The Betrayal of the American Man.* New York: William Morrow.

Faris, J.C. 1979. *Cat Harbour: A Newfoundland Fishing Settlement.* St. John's: ISER Books.

Felt. L. 1987. *Take the "Bloods of Bitches" to the Gallows: Cultural and Structural Constraints Upon Interpersonal Violence in Rural Newfoundland.* St. John's: Institute of Social and Economic Research.

Felt, L., and P.R. Sinclair. 1992. "Everybody Does It: Unpaid Work in a Rural Peripheral Region," *Work, Employment and Society* 6, 1: 43-64.

Ferguson, H. 2001. "Men and Masculinities in Late-modern Ireland," in B. Pease and K. Pringle, eds., *A Man's World? Changing Men's Practices in a Globalised World.* London: Zed Books.

Ferguson, M.E. 1996. "Making Fish – Salt-Cod Processing on the East Coast of Newfoundland – A Study in Historic Occupational Folklife," M.A. thesis, Memorial University of Newfoundland.

Fine, M., L. Weis, J. Addelston, and J. Marusza. 1997. "(In)Secure Time: Constructing White Working-class Masculinities in the Late 20th Century," *Gender & Society* 11, 1: 52-68.

Finlayson, A.C. 1994. *Fishing for Truth: A Sociological Analysis of Northern Cod Stock Assessment from 1977-1990.* St. John's: ISER Books.

Fish, Food and Allied Workers. Various years. *Union Advocate* 1, 2 (Spring/Summer 1992); 3, 1 (Fall/Winter 1994); 3, 1 (Spring/Summer 1994); 4, 2 (Spring/Summer 1996); 5, 1 (Winter 1997). St. John's: FFAW.

———. 2003. "E.I. Act punished seasonal workers." At: www.ffaw.nf.ca/factsheet3.html

Fishery Research Group. 1986. *The Social Impact of Technological Change in Newfoundland's Deepsea Fishery.* St. John's: Institute of Social and Economic Research.

Forrest, D. 2002. "Lenin, the Pinguero, and Cuban Imaginings of Maleness in Times of Scarcity," in F. Cleaver, ed., *Masculinities Matter! Men, Gender and Development*. London: Zed Books.

Gerrard, S. 1995. "When Women Take the Lead: Changing Conditions for Women's Activities, Roles and Knowledge in North Norwegian Fishing Communities," *Social Science Information* 34, 4: 593-631.

Government of Newfoundland and Labrador. 2003. Selected Economic Indicators: Newfoundland and Labrador, 1992 to 2002. At: www.stats.gov.nl.ca/flashsheets/SelEconIndicators.PDF

Gradman, T.J. 1994. "Masculine Identity from Work to Retirement," in E.H. Thompson Jr., ed., *Older Men's Lives*. London: Sage.

Grzetic, Brenda. 2002. "Between Life and Death: Women Fish Harvesters in Newfoundland and Labrador," M.A. thesis, Memorial University of Newfoundland.

———. 2003. "Women Fish Harvesters in Newfoundland and Labrador: Their Work, Learning and Health." At: www.safetynet.mun.ca/pdfs/Brenda%20Grzetic.pdf

Harding, S. 1991. *Whose Science? Whose Knowledge? Thinking from Women's Lives*. Ithaca, N.Y.: Cornell University Press.

Harper, S. 1994. "Subordinating Masculinities/Racializing Masculinities: Writing White Supremacist Discourse on Men's Bodies," *Masculinities* 2, 4: 1-20.

Harris, L. 1990. *Independent Review of the State of the Northern Cod Stock*. Ottawa: Department of Fisheries and Oceans.

Hearn, J., and D.L. Collinson. 1994. "Theorizing Unities and Differences between Men and between Masculinities," in Brod and Kaufman (1994).

Hondagneu-Sotelo, P., and M.A. Messner. 1994. "Gender Displays and Men's Power: The 'New Man' and the Mexican Immigrant Man," in Brod and Kaufman (1994).

Human Resources Development Canada. 1998. *Evaluation of The Atlantic Groundfish Strategy (TAGS): TAGS/HRDC Final Evaluation Report*. Ottawa: Human Resources Development Canada.

Hutchings, J.A., and R.A. Myers. 1995. "The Biological Collapse of Atlantic Cod off Newfoundland and Labrador: An Exploration of Historical Changes in Exploitation, Harvesting Technology, and Management," in R. Arnason and L. Felt, eds., *The North Atlantic Fisheries: Successes, Failures and Challenges*. Charlottetown: Institute of Island Studies.

Judd, C., E. Smith, and L. Kidder. 1991. *Research Methods in Social Relations*, 6th ed. Fort Worth: Harcourt Brace Jovanovich.

Kabeer, N. 1995. *Reversed Realities: Gender Hierarchies in Development Thought*. London: Verso.

Kaufman, M. 1994. "Men, Feminism, and Men's Contradictory Experiences of Power," in Brod and Kaufman (1994).

Keith, P.M. 1994. "A Typology of Orientations toward Household and Marital Roles of Older Men and Women," in E.H. Thompson Jr., ed., *Older Men's Lives.* London: Sage.

Kimmel, M.S. 1994. "Masculinity as Homophobia: Fear, Shame, and Silence in the Construction of Gender Identity," in Brod and Kaufman (1994).

———. 2001. "Global Masculinities: Restoration and Resistance," in B. Pease and K. Pringle, eds., *A Man's World? Changing Men's Practices in a Globalized World.* London: Zed Books.

——— and M. Kaufman. 1994. "Weekend Warriors: The New Men's Movement," in Brod and Kaufman (1994).

Kloppenburg, J. 1991. "Social Theory and the De/Reconstruction of Agricultural Science: Local Knowledge for an Alternative Agriculture," *Rural Sociology* 56, 4: 519-48.

Kuhn, T. 1974. *The Structure of Scientific Revolutions.* Chicago: University of Chicago Press.

Leach, B. 1997. "The New Right and the Politics of Work and Family in Hamilton," *Atlantis* 21, 2 (Spring): 35-46.

———. 2000. "Transforming Rural Livelihoods: Gender, Work, and Restructuring in Three Ontario Communities," in S.M. Neysmith, ed., *Restructuring Caring Labour: Discourse, State Practice, and Everyday Life.* Toronto: Oxford University Press.

——— and A. Winson. 1995. "Bringing 'Globalization' Down to Earth: Restructuring and Labour in Rural Communities," *Canadian Review of Sociology and Anthropology* 31, 3: 341-64.

Luxton, M. 1998. "Families and the Labour Market: Coping Strategies from a Sociological Perspective," in *How Families Cope and Why Policymakers Need to Know,* CPRN Study No. F/02. Ottawa: Canadian Policy Research Networks.

———. 2001. "Family Responsibilities: The Politics of Love and Care," 32nd Annual Sorokin Lecture, University of Saskatchewan, Saskatoon.

——— and J. Corman. 2001. *Getting By in Hard Times: Gendered Labour at Home and at Work.* Toronto: University of Toronto Press.

Mac An Ghaill, M. 1994. "The Making of Black English Masculinities," in Brod and Kaufman (1994).

MacDonald, M., and P. Connelly. 1990a. "A Leaner Meaner Industry: A Case Study of 'Restructuring' in the Nova Scotia Fishery," in B. Fairley, C. Leys, and J. Sacouman, eds., *Restructuring and Resistance: Perspectives from Atlantic Canada.* Toronto: Garamond Press.

——— and ———. 1990b. "Class and Gender in Nova Scotia Fishing Communities," in Fairley, Leys, and Sacouman, eds., *Restructuring and Resistance.*

Margold, J.A. 1994. "Global Disassembly: Migrant Masculinity in the Transnational Workplace," *Masculinities* 2, 3: 18-36.

McCay, B.J. 1988. "Fish Guts, Hair Nets and Unemployment Stamps: Women and Work in Co-operative Fish Plants," in P.R. Sinclair, ed., *A Question of Survival: The Fisheries and Newfoundland Society.* St. John's: ISER Books.

———. 1999. "'That's Not Right': Resistance to Enclosure in a Newfoundland Crab Fishery," in D. Newell and R.E. Ommer, eds., *Fishing Places, Fishing People: Traditions and Issues in Canadian Small-Scale Fisheries.* Toronto: University of Toronto Press.

———. 2000. "Markets, Communities and Ecosystems in Fisheries Management," paper presented at Gender, Globalization and Fisheries Workshop, Memorial University of Newfoundland, May.

McGrath, C., B. Neis, and M. Porter, eds. 1995. *Their Lives and Times: Women in Newfoundland and Labrador, A Collage.* St. John's: Killick Press.

Merchant, C. 1980. *The Death of Nature: Women, Ecology and the Scientific Revolution.* San Francisco: Harper and Row.

Morris, L.D. 1984. "Redundancy and Patterns of Household Finance," *Sociological Review* 32, 3: 492-523.

———. 1991. "Locality studies and the household," *Environment and Planning A* 23: 165-77.

Moser, C. 1993. *Gender Planning and Development: Theory, Practice and Training.* London: Routledge.

Mulkay, M. 1979. *Science and the Sociology of Knowledge.* London: George Allen and Unwin.

Munk-Madsen, E. 1996. "Wife the Deckhand; Husband the Skipper: Authority and Dignity among Fishing Couples," paper presented at the Research Course "Global Resource Crises – Marginalised Men and Women," Norwegian College of Fisheries Science, 5-9 Sept.

———. 1998. "The Norwegian Fishing Quota System: Another Patriarchal Construction?" *Society and Natural Resources* 11: 229-40.

Murray, M., D. Fitzpatrick, and C. O'Connell. 1997. "Fishermen's Blues: Factors related to accidents and safety among Newfoundland fishermen," *Work and Stress* 11, 3: 292-7.

Nadel-Klein, J., and D.L. Davis, eds. 1988. *To Work and To Weep: Women in Fishing Economies.* St. John's: ISER Books.

Neis, B. 1991. "Flexible Specialization: What's that got to do with the price of fish?" *Studies in Political Economy* 36: 145-76.

———. 1992. "Fishers' Ecological Knowledge and Stock Assessment in Newfoundland and Labrador," *Newfoundland Studies* 8, 2: 155-75.

———. 1993. "From 'Shipped Girls' to 'Brides of the State': The Transition from Familial to Social Patriarchy in the Newfoundland Fishing Industry," *Canadian Journal of Regional Science* 16, 2: 185-211.

———. 2001. "We've Talked: Restructuring and Women's Health in Rural Newfoundland: Final Report to Health Canada Population Health Fund." At: www.coastsunderstress.ca/pubs/weve_talked_report_nov_2001.pdf

———, B. Grzetic, and M. Pidgeon. 2001. "From Fishplant to Nickel Smelter: Health Determinants and the Health of Newfoundland's Women Fish and Shellfish Processors in an Environment of Restructuring." St. John's: Memorial University of Newfoundland.

——— and L. Felt, eds. 2000. *Finding Our Sea Legs: Linking Fishery People and Their Knowledge with Science and Management.* St. John's: ISER Books.

——— and S. Williams. 1993. *Occupational Stress and Repetitive Strain Injuries: Research Review and Pilot Study.* St. John's: Institute of Social and Economic Research.

——— and ———. 1997. "The New Right, Gender and the Fisheries Crisis: Local and Global Dimensions," *Atlantis* 21, 2: 47-62.

Nemec, T.F. 1972. "I Fish with My Brother: The Structure and Behaviour of Agnatic-Based Fishing Crews in a Newfoundland Irish Outport," in R. Andersen and C. Wadel, eds., *North Atlantic Fishermen: Anthropological Essays on Modern Fishing.* St. John's: ISER Books.

Newfoundland and Labrador Statistics Agency. 2002. Population by Age Group, Newfoundland and Labrador, July 1, 1971 to 2002. Government of Newfoundland and Labrador. At: http://www.nfstats.gov.nf.ca/Statistics/Population/PDF/PopAgeSex_BS.PDF

———. 2003. Statistics: Labour Market. Government of Newfoundland and Labrador. At: www.stats.gov.nl.ca/statistics/labour/unemployment_rate.asp

Neysmith, S.M. 2000. "Networking Across Difference: Connecting Restructuring and Caring Labour," in S.M. Neysmith, ed., *Restructuring Caring Labour: Discourse, State Practice, and Everyday Life.* Toronto: Oxford University Press.

Ommer, R.E. 1995. "Policy, Fisheries Management and Development in Rural Newfoundland," in A. Gilg, ed., *Progress in Rural Policy and Planning.* Chichester: John Wiley & Sons.

———. 1998. *Final Report of the Eco-Research Project: Sustainability in a Changing Cold-Ocean Coastal Environment.* St. John's: Institute of Social and Economic Research.

Palsson, G. 2000. "Finding One's Sea Legs: Learning, the Process of Enskilment, and Integrating Fishers and Their Knowledge into Fisheries Science and Management," in Neis and Felt (2000).

Pease, B., and K. Pringle. 2001. "Introduction: Studying Men's Practices and Gender Relations in a Global Context," in B. Pease and K. Pringle, eds., *A Man's World? Changing Men's Practices in a Globalized World.* London: Zed Books.

Pollert, A. 1981. *Girls, Wives, Factory Lives.* London: Macmillan.

Porter, M. 1993. *Place and Persistence in the Lives of Newfoundland Women.* Aldershot: Avebury.

Power, N. 1997. "Women, Processing Industries and the Environment: A Sociological Analysis of Women Fish and Crab Processing Workers' Local Ecological Knowledge," M.A. thesis, Memorial University of Newfoundland.

———. 2000. "Women Processing Workers as Knowledgeable Resource Users: Connecting Gender, Local Knowledge and Development in the Newfoundland Fishery," in Neis and Felt (2000).

Pringle, K., and B. Pease. 2001. "Afterword: A Man's World? Rethinking Commonality and Diversity in Men's Practices," in B. Pease and K. Pringle, eds., *A Man's World? Changing Men's Practices in a Globalized World.* London: Zed Books.

Robinson, J. 1995. "Women and Fish Plant Closure: The Case of Trepassey, Newfoundland," in McGrath et al. (1995).

Rowe, A. 1991. *Effect of the Crisis in the Newfoundland Fishery on Women Who Work in the Industry.* St. John's: Women's Policy Office, Government of Newfoundland and Labrador.

Silk, V. 1995. "Women and the Fishery," in McGrath et al. (1995).

Sinclair, P.R. 1987. *State Intervention and the Newfoundland Fisheries.* Aldershot: Avebury.

———. 1988a. "Introduction," in P.R. Sinclair, ed., *A Question of Survival: The Fisheries and Newfoundland Society.* St. John's: ISER Books.

———. 1988b. "The State Encloses the Commons: Fisheries Management from the 200-Mile Limit to Factory Freezer Trawlers," in Sinclair, ed., *A Question of Survival.*

———. 2002. "Leaving and Staying: Bonavista Residents Adjust to the Moratorium," in R. Ommer, ed., *The Resilient Outport: Ecology, Economy, and Society in Rural Newfoundland.* St. John's: ISER Books.

——— and L. Felt. 1992. "Separate Worlds: Gender and Domestic Labour in an Isolated Fishing Region," *Canadian Review of Sociology and Anthropology* 29, 1: 55-71.

———, H. Squires, and L. Downton. 1999. "A Future without Fish? Constructing Social Life on Newfoundland's Bonavista Peninsula after the Cod Moratorium," in D. Newell and R. Ommer, eds., *Fishing Places, Fishing People: Traditions and Issues in Canadian Small-Scale Fisheries.* Toronto: University of Toronto Press.

Skaptadottir, U. D., and R. H. Proppe. 2000. "The application of globalization and feminist theory to an analysis of changing property regimes in Icelandic fisheries," paper presented at the Workshop on Gender, Globalization and Fisheries, Memorial University of Newfoundland, 6-12 May.

Skipton, S. 1997. "Cuts, Privatization and Deregulation: A Look at the Impact of Some New Right Government Policies on Women," *Atlantis* 21, 2 (Spring/Summer): 17-27.

Solomon, K., and P.A. Szwabo. 1994. "The Work-oriented Culture: Success and Power in Elderly Men," in E.H. Thompson Jr., ed., *Older Men's Lives*. London: Sage.

Statistics Canada. 1996. Community Profiles. At: http://www12.statcan.ca/english/Profil/PlaceSearchForm1.cfm

———. 2001. Community Profiles. At: http://www12.statcan.ca/english/Profil01/PlaceSearchForm1.cfm

———. 2003a. *The Daily*, 24 Jan. 2003. "Year-end Labour Market Review 2002." At: http://www.statcan.ca/Daily/English/030124/ d030124a.htm

———. 2003b. A Profile of the Canadian Population: Where We Live. At: http://geodepot.statcan.ca/Diss/Highlights/Index_e.cfm

Stiles, R.G. 1972. "Fishermen, Wives and Radios: Aspects of Communication in a Newfoundland Fishing Community," in R. Andersen and C. Wadel, eds., *North Atlantic Fishermen: Anthropological Essays on Modern Fishing*. St. John's: ISER Books.

Szala, K. 1978. "Clean Women and Quiet Men: Courtship and Marriage in a Newfoundland Fishing Village," M.A. thesis, Memorial University of Newfoundland.

Szwed, J. 1966. *Private Cultures and Public Imagery: Interpersonal Relations in a Newfoundland Peasant Society*. St. John's: ISER Books.

Thompson, E.H., Jr. 1994. "Older Men as Invisible Men in Contemporary Society," in E.H. Thompson Jr., ed., *Older Men's Lives*. London: Sage.

Tolson, A. 1977. *The Limits of Masculinity*. London: Tavistock.

Veltmeyer, H. 1990. "The Restructuring of Capital and the Regional Problem," in B. Fairley, C. Leys, and J. Sacouman, eds., *Restructuring and Resistance: Perspectives from Atlantic Canada*. Toronto: Garamond Press.

Wadel, C. 1969. *Marginal Adaptations and Modernization in Newfoundland: A Study of Strategies and Implications in the Resettlement and Redevelopment of Outport Fishing Communities*. St. John's: ISER Books.

———. 1973. *Now, Whose Fault Is That? The Struggle for Self-Esteem in the Face of Chronic Unemployment*. St. John's: ISER Books.

Walker, K. 1994. "'I'm Not Friends the Way She's Friends': Ideological and Behavioral Constructions of Masculinity of Men's Friendships," *Masculinities* 2, 2: 38-55.

White, S. 1997. "Men, masculinities, and the politics of development," *Gender and Development* 5, 2 (June): 14-22.

Williams, S. 1996. *Our Lives Are at Stake: Women and the Fishery Crisis in Newfoundland and Labrador.* St. John's: Institute of Social and Economic Research.

Willis, P.E. 1977. *Learning to Labour: How Working Class Kids get Working Class Jobs.* Farnborough: Saxon House.

Women in Resource Development Committee. 2003. "Fishing Industry: Overview and Trends." http://www.wrdc.nf.ca/fisheries.htm

Wright, M. 1995. "Women, Men and the Modern Fishery: Images of Gender in Government Plans for the Canadian Atlantic Fisheries," in McGrath et al. (1995).

Wright, M. 2001. *A Fishery for Modern Times: The State and the Industrialization of the Newfoundland Fishery, 1934-1968.* Toronto: Oxford University Press.

Yarrow, M. 1992. "Class and Gender in the Developing Consciousness of Appalachian Coal-Miners," in A. Sturdy, D. Knights, and H. Willmott, eds., *Skill and Consent: Contemporary Studies in the Labour Process.* London: Routledge.

# Index

ISER BOOKS

*Studies*

69 **What Do They Call a Fisherman?** – Nicole Gerarda Power
68 **Narratives at Work: Women, Men, Unionization, and the Fashioning of Identities** – Linda Kathleen Cullum
67 **A Way of Life That Does Not Exist: Canada and the Extinguishment of the Innu** – Colin Samson
66 **Enclosing the Commons: Individual Transferable Quotas in the Nova Scotia Fishery** – Richard Apostle, Bonnie McCay, and Knut H. Mikalsen
65 **Cows Don't Know It's Sunday: Agricultural Life in St. John's** – Hilda Chaulk Murray
64 **Place Names of the Northern Peninsula** - Edited by Robert Hollett and William J. Kirwin
63 **Remembering the Years of My Life: Journeys of a Labrador Inuit Hunter** - recounted by Paulus Maggo, edited with an Introduction by Carol Brice-Bennett
62 **Inuit Morality Play: The Emotional Education of a Three-Year-Old** - Jean Briggs
61 **The Marke of Power: Helgeland and the Politics of Omnipotence** - George Park
60 **Literacy for Living: A Study of Literacy and Cultural Context in a Rural Canadian Community** - William T. Fagan
59 **A Time of Reckoning: The Politics of Discourse in Rural Ireland** - Adrian Peace
58 **The People of Sheshatshit: In the Land of the Innu** - José Mailhot
57 **Looking Out for the Lads: Community Action and the Provision of Youth Services in an Urban Irish Parish** - Stephen A. Gaetz
56 **Making a World of Difference: Essays on Tourism, Culture and Development in Newfoundland** - James Overton
55 **Quest for Equity: Norway and the Saami Challenge** - Trond Thuen
54 **Rough Food: The Seasons of Subsistence in Northern Newfoundland** - John T. Omohundro
53 **Voices from Off Shore: Narratives of Risk and Danger in the Nova Scotian Deep-Sea Fishery** - Marian Binkley
52 **Fishing for Truth: A Sociological Analysis of Northern Cod Stock Assessments from 1977 to 1990** - Alan Christopher Finlayson
51 **Predictions Under Uncertainty: Fish Assemblage and Food Webs on the Grand Banks of Newfoundland** - Manuel Do Carmo Gomes
50 **Dangling Lines: The Fisheries Crisis and the Future of Coastal Communities: The Norwegian Experience** - Svein Jentoft
49 **Port O'Call: Memories of the Portuguese White Fleet in St. John's, Newfoundland** - Priscilla A. Doel
48 **Sanctuary Denied: Refugees from the Third Reich and Newfoundland Immigration Policy, 1906–1949** - Gerhard P. Bassler
47 **Violence and Public Anxiety: A Canadian Case** - Elliott Leyton, William O'Grady and James Overton

46 **What is the Indian 'Problem': Tutelage and Resistance in Canadian Indian Administration** - Noel Dyck
45 **Strange Terrain: The Fairy World in Newfoundland** - Barbara Rieti
44 **Midwives in Passage: The Modernisation of Maternity Care** - Cecilia Benoit
43 **Dire Straits: The Dilemmas of a Fishery, The Case of Digby Neck and the Islands** - Anthony Davis
42 **Saying Isn't Believing: Conversation, Narrative and the Discourse of Belief in a French Newfoundland Community** - Gary R. Butler
41 **A Place in the Sun: Shetland and Oil - Myths and Realities** - Jonathan Wills
40 **The Native Game: Settler Perceptions of Indian/Settler Relations in Central Labrador** - Evelyn Plaice
39 **The Northern Route: An Ethnography of Refugee Experiences** - Lisa Gilad
38 **Hostage to Fortune: Bantry Bay and the Encounter with Gulf Oil** - Chris Eipper
37 **Language and Poverty: The Persistence of Scottish Gaelic in Eastern Canada** - Gilbert Foster
36 **A Public Nuisance: A History of the Mummers Troupe** - Chris Brookes
35 **Listen While I Tell You: A Story of the Jews of St. John's, Newfoundland** - Alison Kahn
34 **Talking Violence: An Anthropological Interpretation of Conversation in the City** - Nigel Rapport
33 **"To Each His Own": William Coaker and the Fishermen's Protective Union in Newfoundland Politics, 1908-1925** - Ian D.H. McDonald, edited by J.K Hiller
32 **Sea Change: A Shetland Society, 1970-79** - Reginald Byron
31 **From Traps to Draggers: Domestic Commodity Production in Northwest Newfoundland, 1850-1982** - Peter Sinclair
30 **The Challenge of Oil: Newfoundland's Quest for Controlled Development** - J.D. House
28 **Blood and Nerves: An Ethnographic Focus on Menopause** - Dona Lee Davis
27 **Holding the Line: Ethnic Boundaries in a Northern Labrador Community** - John Kennedy
26 **'Power Begins at the Cod End': The Newfoundland Trawlermen's Strike, 1974-75** - David Macdonald
25 **Terranova: The Ethos and Luck of Deep-Sea Fishermen** - Joseba Zulaika
24 **"Bloody Decks and a Bumper Crop": The Rhetoric of Sealing Counter-Protest** - Cynthia Lamson
23 **Bringing Home Animals: Religious Ideology and Mode of Production of the Mistassini Cree Hunters** - Adrian Tanner
22 **Bureaucracy and World View: Studies in the Logic of Official Interpretation** - Don Handelman and Elliott Leyton

20 **You Never Know What They Might Do: Mental Illness in Outport Newfoundland** - Paul S. Dinham
19 **The Decay of Trade: An Economic History of the Newfoundland Saltfish Trade, 1935–1965** - David Alexander
18 **Manpower and Educational Development in Newfoundland** - S.S. Mensinkai and M.Q. Dalvi
17 **Ancient People of Port au Choix: The Excavation of an Archaic Indian Cemetery in Newfoundland** - James A. Tuck
16 **Cain's Land Revisited: Culture Change in Central Labrador, 1775–1972** - David Zimmerly
15 **The One Blood: Kinship and Class in an Irish Village** - Elliott Leyton
14 **The Management of Myths: The Politics of Legitimation in a Newfoundland Community** - A.P. Cohen
12 **Hunters in the Barrens: The Naskapi on the Edge of the White Man's World** - Georg Henriksen
11 **Now, Whose Fault is That? The Struggle for Self-Esteem in the Face of Chronic Unemployment** - Cato Wadel
10 **Craftsman-Client Contracts: Interpersonal Relations in a Newfoundland Fishing Community** - Louis Chiaramonte
9 **Newfoundland Fishermen in the Age of Industry: A Sociology of Economic Dualism** - Ottar Brox
8 **Public Policy and Community Protest: The Fogo Case** - Robert L. DeWitt
7 **Marginal Adaptations and Modernization in Newfoundland: A Study of Strategies and Implications of Resettlement and Redevelopment of Outport Fishing Communities** - Cato Wadel
6 **Communities in Decline: An Examination of Household Resettlement in Newfoundland** - N. Iverson and D. Ralph Matthews
5 **Brothers and Rivals: Patrilocality in Savage Cove** - Melvin Firestone
4 **Makkovik: Eskimos and Settlers in a Labrador Community** - Shmuel Ben-Dor
3 **Cat Harbour: A Newfoundland Fishing Settlement** - James C. Faris
2 **Private Cultures and Public Imagery: Interpersonal Relations in a Newfoundland Peasant Society** - John F. Szwed
1 **Fisherman, Logger, Merchant, Miner: Social Change and Industrialism in Three Newfoundland Communities** - Tom Philbrook

*Papers*

25 **The Resilient Outport: Ecology, Economy, and Society in Rural Newfoundland** - Edited by Rosemary E, Ommer
24 **Finding Our Sea Legs: Linking Fishery People and Their Knowledge with Science and Management** - Edited by Barbara Neis and Lawrence Felt
23 **Just Fish: Ethics and Canadian Marine Fisheries** - Harold Coward, Rosemary Ommer, and Tony Pitcher (eds.)
22 **Labour and Working-Class History in Atlantic Canada: A Reader** - David Frank and Gregory S. Kealey (eds.)

21 **Living on the Edge: The Great Northern Peninsula of Newfoundland** - Lawrence F. Felt and Peter R. Sinclair (eds.)
20 **Pursuing Equality: Historical Perspectives on Women in Newfoundland and Labrador** - Linda Kealey (ed.)
19 **Living in a Material World: Canadian and American Approaches to Material Culture** - Gerald L. Pocius (ed.)
18 **To Work and to Weep: Women in Fishing Economies** - Jane Nadel-Klein and Dona Lee Davis (eds.)
17 **A Question of Survival: The Fisheries and Newfoundland Society** - Peter R. Sinclair (ed.)
16 **Fish Versus Oil: Resources and Rural Development in North Atlantic Societies** - J.D. House (ed.)
15 **Advocacy and Anthropology: First Encounters** - Robert Paine (ed.)
14 **Indigenous Peoples and the Nation-State: Fourth World Politics in Canada, Australia and Norway** - Noel Dyck (ed.)
13 **Minorities and Mother Country Imagery** - Gerald Gold (ed.)
12 **The Politics of Indianness: Case Studies of Native Ethnopolitics in Canada** - Adrian Tanner (ed.)
11 **Belonging: Identity and Social Organisation in British Rural Cultures** - Anthony P. Cohen (ed.)
10 **Politically Speaking: Cross-Cultural Studies of Rhetoric** - Robert Paine (ed.)
9 **A House Divided? Anthropological Studies of Factionalism** - M. Silverman and R.F. Salisbury (eds.)
8 **The Peopling of Newfoundland: Essays in Historical Geography** - John J. Mannion (ed.)
7 **The White Arctic: Anthropological Essays on Tutelage and Ethnicity** - Robert Paine (ed.)
6 **Consequences of Offshore Oil and Gas - Norway, Scotland and Newfoundland** - M.J. Scarlett (ed.)
5 **North Atlantic Fishermen: Anthropological Essays on Modern Fishing** - Raoul Andersen and Cato Wadel (eds.)
3 **The Compact: Selected Dimensions of Friendship** - Elliott Leyton (ed.)
2 **Patrons and Brokers in the East Arctic** - Robert Paine (ed.)
1 **Viewpoints on Communities in Crisis** - Michael L. Skolnik (ed.)

ISER Books
Faculty of Arts Publications
Memorial University
St. John's, NL
A1C 5S7

Telephone: (709) 737-7474 FAX : (709) 737-7560
email: iser-books@mun.ca
WEB site: http://www.mun.ca/iser/